AF598261

Steam Turbine Governing and Overspeed Protection

Steam Plant Committee

J K Clayton
Hopkinsons Limited

I Clifford
Enrion Power Operations Limited

A Davenport
Nooter/Eriksen Limited

E W East
Edward East Limited

N G Foster
Energy Efficiency Office

J Grant (Past Chair)
Siemens Power Gen Systems Limited

J E Jesson (Chair)
Mitsui Babcock Energy Limited

M Macrae
Mitsui Babcock Energy Limited

R J Martin
National Power plc

J Muscroft
GEC Alsthom

F Nash
McLellan and Partners Limited

A G Sheard
Allen Steam Turbines

S Simpson
PowerGen plc

C J Waterman
Magnox Electric

P Winkle
Scottish Nuclear Limited

Seminar Adviser

R J Martin CEng, MIMechE
Turbine Generator Group
National Power plc
Swindon, UK

Papers presented at a one-day seminar *Steam Turbine Governing and Overspeed Protection* held at IMechE Headquarters, London, UK, on 28 May 1998.

IMechE
Seminar Publication

Steam Turbine Governing and Overspeed Protection

Organized by the Steam Plant Committee of
The Institution of Mechanical Engineers (IMechE)

IMechE Seminar Publication 1998–10

Published by Professional Engineering Publishing Limited for the Institution of Mechanical Engineers, Bury St Edmunds and London, UK.

First Published 1998

ISBN 1 86058 162 5

A CIP catalogue record for this book is available from the British Library.

Printed by The Ipswich Book Company, Suffolk, UK.

Contents

Related Titles of Interest

Title	**Editor/Author**	**ISBN**
Handbook of Mechanical Works Inspection	Clifford Matthews	1 86058 047 5
Advanced Steam Plant	IMechE Conference 1997-2	1 86058 097 1
Remanent Life Prediction	IMechE Seminar 1998–4	1 86058 154 4
Gas Explosions in CCGT and Steam Plants	IMechE Seminar 1998–3	1 86058 169 2
Plant Monitoring and Maintenance Routines	IMechE Seminar 1998–2	1 86058 087 4
Aerodynamics of Turbomachinery	IMechE Seminar 1996–21	1 86058 051 3

For the full range of titles published by Professional Engineering Publishing contact:

Sales Department
Professional Engineering Publishing Limited
Northgate Avenue
Bury St Edmunds
Suffolk
IP32 6BW
UK

Tel: 01284 724384
Fax: 01284 718692

S548/001/98

Frequency control on the UK system – requirements and dispatch

J PHILLIPS BEng, AMIEE
The National Grid Company plc, Wokingham, UK

INTRODUCTION

The Electricity Supply Regulations state that the system frequency should be maintained at 50 Hz ± 1% save in exceptional circumstances. The Transmission Licence places an obligation on the National Grid Company [NGC] to plan and operate the system to ensure compliance with the Electricity Supply Regulations.

NGC SYSTEM RESPONSE REQUIREMENTS

In order to maintain the frequency within the above standards, NGC must be able to cater for large, unexpected losses in generation and demand which can occur at any time without notice.

NGC must be able to contain the frequency to the following levels:

- Up ± 0.2 Hz — maximum frequency change for losses up to ± 300 MW;
- ± 0.5 Hz — maximum frequency change for losses greater than ± 300 MW and less than or equal to ± 1000 MW;
- - 0.8 Hz — maximum frequency change for losses greater than 1000 MW and less than or equal to 1320 MW – controlled to 49.5 Hz within 1 minute.

In order to maintain the system frequency to these levels, NGC carries response.

RESPONSE

Response can be provided by Generation running at part load, able to provide an automatic change in output within set timescales, or demand tripping on low frequency relays at preset levels.

For the purposes of this paper, response provided by generators only will be considered.

In order to control the frequency following a large loss, the drop must first be contained, then sufficient energy must be injected into the system such that the frequency will recover. These different requirements are set out below:

i) Primary response: this is an automatic increase in output which occurs following a drop in frequency, which is fully available within 10 seconds and is sustainable for 30 seconds.

ii) Secondary response: this is an automatic increase in output which occurs following a drop in frequency, which is fully available within 30 seconds and is sustainable for 30 minutes.

iii)High frequency response: this is an automatic decrease in output which is fully available within 30 seconds and is sustainable indefinitely.

Primary response is used to contain a frequency fall; it can be shown that for large losses on the UK system, the time taken for the frequency to reach its lowest point is 10 seconds, hence the timescale involved in the specification of primary response.

Secondary response is used to recover the frequency. After a large loss, other plant will be manually instructed up in order to replace the lost MWs, so secondary response must be capable of being sustained until the replacement plant has reached its full output, typically 30 minutes.

High frequency response is required to cater for demand losses and islanding situations.

RESPONSE REQUIREMENTS

To calculate the amount of response which is required on the UK system, NGC has a dynamic model of the system which it uses to simulate the effect of losses. The model uses information on genset characteristics, such as governor parameters and various characteristics of the turbo-alternator, which have been provided under the Grid Code.

The simulation is run such that a specific loss is applied to the system when it is at a particular demand, and the amount of response required to contain that loss to the relevant frequency deviation is calculated.

This simulation is repeated for the full range of system demands, and a range of losses.

From these results, a set of curves are produced. These give the amount of response required for a particular loss against system demand; this allows the actual requirement to be calculated on a continuous basis. A typical curve can be seen in Figure 1.

RESPONSE CONTRACTS

All generators are required by the Grid Code to have the capability to provide response. The Grid Code capability of each genset is set down in an Ancillary Services Contract. This contract gives information on the amount of primary, secondary and high frequency response given by the genset for a number of different frequency deviations, over a range of loading levels.

The contract also includes a price which is paid to the generators when they are allocated to provide response.

Information on the contracted levels of response is loaded into a database, for use in real time systems.

RESPONSE HOLDING IN REAL TIME

Assessments are done at various timescales in order to ensure that there is likely to be sufficient plant on the system to meet demand and to provide frequency response.

At the week ahead stage, an indicative system response level is published by NGC in a Weekly Operational Policy, which is sent to each generator.

At the day ahead stage, an assessment is done of the requirements, based on predicted demand at the peaks and troughs of the following day. At this stage, there is a knowledge of the generating units which are likely to be running, and a check is made to ensure there is sufficient plant available to meet demand and System response requirements.

In the Control phase, an online tool is used to determine the most appropriate plant to carry response in real time. This tool contains a complex model, with a linear program, which has information including the response requirements curves, the contractual capability of each genset on the system, the actual and predicted demands over the next 2 hours and the different risks on the system.

In order to provide response, a generating unit must be deloaded, and so may incur constrained off payments through the Pool; this is one part of the cost of holding response.

The second part of the cost is the actual Ancillary Services contract cost, which covers operating the generator in frequency sensitive mode (i.e. to cover throttling losses and other wear and tear).

The online tool calculates the most economic way of meeting the response requirements, and this advice is used by NGC's Generation Despatch Engineers who give the appropriate loading instructions to generating units, along with instructions to provide response.

RESPONSE SHORTFALL

There are some situations under which it can be difficult to both meet demand and to hold sufficient frequency response to cover the risks on the system. The most common situation occurs during the summer months, when system demand is at its lowest.

It has been seen from Figure 1 that the response requirement is at its highest when the system demand is at its lowest.

However, under these conditions, a substantial proportion of plant on the system may be inflexible, i.e. unwilling or unable to deload in order to provide frequency response.

Under these circumstances, NGC may be forced to use Grid Code provisions to desynchronise inflexible plant in order to ensure that there is enough flexible plant on the system to meet demand and the response requirements so as to maintain security.

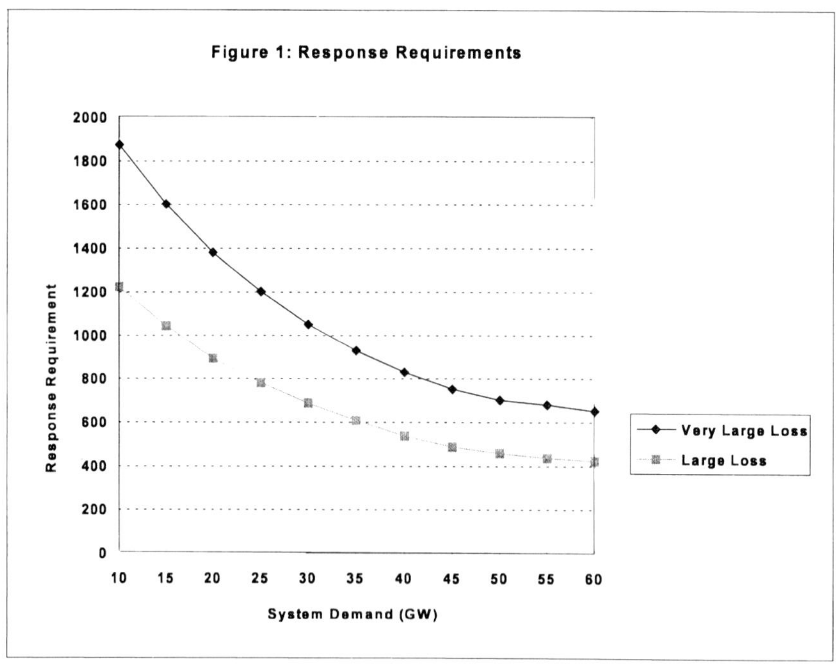

S548/002/98

Impact of market changes upon power plant control

A W OGLE CEng, MIMechE and **M RENDER** BSc, MEng
GEC ALSTHOM Power Generation Limited, Rugby, UK

1. INTRODUCTION

The main evolution of the steam turbine governor has occurred during the past 30 years. Governors are now extremely sophisticated and incorporate a wide variety of control and interfacing functions. We will demonstrate that on all fronts the changes which have taken place in power generation during the last 8 to 9 years have exploited these features in order to establish plant which can perform more effectively and efficiently.

2. HISTORY

In the beginning there was the flyball mechanical governor which in its earliest form had direct action upon the steam valve (Figure 1). This form of device served the industry well for many years and whilst the complexity of the steam valve actuator system and signal transmission became more sophisticated utilising mechanical hydraulic power amplifiers the basic governor head remained essentially unchallenged until the mid-1960's. Around this time electronic control systems became a possibility and were introduced.

Although the electronic governing system for the modern large steam turbine has changed significantly over the past twenty five years its principal role has not. The governor is still required to control the turbine under the same fundamental conditions:-

1. Start up:- control of machine speed whilst starting the turbine, especially important during synchronisation of the unit to the grid system.

2. Negotiation of Severe Transients:- control of turbine speed following load rejections when the unit has become separated from the system load.

3. On load operation:- control of system frequency, in conjunction with other units on the system, within acceptable limits of frequency variation in response to the random changes in demand resulting from normal system operation.

The advent of electronic governors gave birth to a rapid increase in the sophistication of governing systems. It was possible to create more complex control configurations and to interface the governing system more easily with operator controls and other closely allied systems.

The use of electronic systems was viewed with suspicion by some users and considerable effort was taken to address the integrity and security of the electronic governor to permit it to match that perceived of the mechanical governor, which was, in the view of many, virtually fail-safe and totally reliable in operation. The mechanical governor had been refined and developed over a hundred years to have failure modes to safety, which were not always possible with certainty with the electronic governor.

The electro-hydraulic governor which evolved obtained very high safety levels by triplicated redundancy of the main speed/load processing and by triplication or duplication of the steam valve control loops. At this time most manufacturers introduced high pressure hydraulics for steam valve actuators for their large turbines, operating with hydraulic pressures in the range 120 to 180 bars. The use of fire resistant fluids for these high pressure servo systems became common.

The electronic governing systems proved to have high degrees of accuracy, speed of response and reliability. Two from three voting of the control signals allowed the governor to continue in operation in the presence of single channel faults and permitted faults occurring to be easily detected and identified.

The flexibility of operation and reliability of such governors was unquestionably very good. However the equipment cost and complication was high and the setting up procedures fairly extensive.

The introduction of microprocessor technology enabled review of the deficiencies and limitations of existing technology. Manufacturers reviewed the possibilities of the new equipment in order to obtain the maximum exploitation for the future control systems. The possibilities of exploitation of the pseudo intelligence of the processor to improve integrity rather than through redundant hardware comparison were realised.

The intelligent processor permitted the implementation of logic necessary to decide whether incoming signals are sensible and whether the process is functioning correctly.

There are two distinct aspects of microprocessor governing equipment, hardware and software. The hardware platform must be chosen having regard for reliability and fail-safe operation.

Software is readily capable of modification. It is mainly for this reason that careful attention must be given to the software structure, it must be robust and modular.

Some manufacturers have employed multi-level structures to permit segregation between high integrity and high availability governing functions which do not change between applications and the less important supervision, co-ordination and management functions which are likely to be influenced by the particular application. (Figure 2).

This has benefit that those governing functions which are vital to the safety and continued operation of the turbine are those which are least likely to change with application. They are generally simple, well defined and demanding of a fast response.

These secondary governing functions, which include various outer control modifier loops, plant test functions, calibration aids, commissioning tools and the high level operator interfaces, are segregated. Loss of these functions does not endanger the turbine and does not force a turbine shutdown.

The advancement in technology, in addition to bringing improvement in quality of control, has provided the ability to interface more conveniently with other control and information centres of the plant, permitting such features as automatic run up and loading control, boiler/turbine control and generation control to be more effectively engineered.

3. IMPACT OF A CHANGING INDUSTRY

During the past 8 to 9 years the dramatic changes in the industry have had very significant effect upon manufacturers of generating plant.

a) The restructuring of the industry and its privatisation in the UK necessitated a document, the Grid Code, which defined operating principles and relationships between NGC and Generators. This resulted in a new for more accurate prediction and implementation of control, together with means of assuring control quality.

b) Privatisation of the supply industry has caused added focus upon cost and efficiency of plant. Initially in the UK market and generally across the world.

c) Changes in UK regulations which permitted natural gas to be used for power generation has had effect upon the type of plant being ordered by generating companies. There has been a dramatic upsurge in both application, complexity and size of combined cycle generating plant.

These imposed requirements upon Generators for both new and existing plant which caused them to review the manner in which their plant was to be operated, and the means by which its performance could be predicted and assessed.

3.1 Retrofit Governors

The high investment cost of new plant has contributed to a substantial increase in plant life extension. In the retrofit market this life extension has been seen both in the UK and overseas in a number of ways. One of these has been the application of retrofit turbine controls aimed at improving the accuracy and predictability of control.

The mechanical hydraulic governor had served the generators well over the years, but its appropriateness to modern power system control left much to be desired. These governors exhibit non linear characteristics which lead to wide variations in incremented droop and variation in characteristics during operational periods between overhauls resulting in unpredictable response characteristics. This presents difficulty for the generators in declaring their reserve capacity and even more difficulty in demonstrating it. To overcome these effects it has become practice to retrofit units with replacement electronic governors.

The cost of such retrofits is often critical. A replacement governor is difficult to substantiate on economic grounds since improvements in quality of control are difficult to relate to cost savings.

Full replacement governors involve costly modifications to hydraulic systems and are usually unnecessary since the existing low pressure hydraulic amplifiers and steam valve actuators demonstrate adequate performance with perfectly acceptable hysteresis.

A governor replacement which involves an electro hydraulic converter for each controlled steam valve is certainly the most costly option. A single electro hydraulic interface provides a cheaper, but less flexible solution. The steam valves retain their fixed relationship with each other and linearisation of steam valve characteristic to compensate for the Rateau factor on multi-valve machines with sequenced HP steam governing valves is not readily applied.

A simpler solution is the provision of a frequency compensated load controller interfacing with the existing speed setting motor or directly with the governed oil system.

It is worth mentioning that the proven reliability of modern electronic equipment, coupled with the demonstrated good performance of electronic governing systems over the years has created user confidence to the extent that single channel systems without redundancy are being accepted for many retrofit projects upon larger units.

4. IMPROVEMENTS TO CONTROL OF FOSSIL FIRED PLANT

4.1 Integration of Boiler and Turbine Controls

Today governor controls are generally integrated into a total boiler/turbine control strategy. Whilst this originates from a desire to implement greater levels of automation to lessen the demands upon the operator and contribute to both availability and component life of the plant it can, if thoughtfully applied, provide additional benefits to efficiency and operational flexibility. Modern plant is required to respond to load changes quickly without any part of the plant entering dangerous or insecure operating states. This has led to increased adoption of co-ordinated control schemes.

Co-ordination is necessary to optimise the operation of the plant. Turbine load is only increased with a parallel input to the boiler controls. Co-ordination should account for the firing inertia of the boiler such that design limits are not exceeded and the boiler must be prevented from causing steam condition transients which could cause damage to the turbine in the short or long term.

Ideally the unit control scheme should monitor the plant state and control the turbine and boiler so as to maximise load and availability and ensure security of continuous operation.

The main functions of co-ordinated control of a modern fossil fired unit are:-

- To start the unit automatically as quickly and efficiently as possible.

- To run the unit to a safe operating point in the case of disturbances with the aim of minimising the occasions when it is necessary to trip the boiler as result of such disturbances.

- To provide unit management and conditions of load variations according to efficiency criteria and to maintain stress levels in both boiler and turbine below acceptable limits.

Co-ordination generally employs the well known principles of Boiler Follows Turbine (BFT) and Turbine Follows Boiler (TFB) control.

One popular practice uses a direct energy balance approach to co-ordination (Figure 3a). This is largely used in North America and markets influenced by that technology.

Pressure and load controllers continuously control the firing and turbine steam admission, providing co-ordination with an adjustable blend of the two basic forms of unit control (TFB & BFT). During load changes the load and pressure controllers influence the turbine governor valves and boiler firing in parallel. The degree of boiler/turbine interaction is determined by the relative weighting of the pressure function which controls the degree to which stored energy is released during a transient.

The load signal includes a frequency compensation component which may be modified to strengthen or weaken the steady state regulation contribution of the plant. The load signal is biased by pressure error and combined with the actual measured MW to create an error which is integrated to provide a turbine load reference.

Although the direct energy balance approach does provide benefit in terms of stability and accuracy with minimum disturbance of system parameters this is at the expense of some of the initial response and maximum use of the boiler stored energy which could be made available for transient increase in generation is prevented. This results in significant loss in generation in the seconds following a disturbance at a critical time when the response of slower units is awaited.

An alternative approach, more popular in Europe has resulted in the adoption of systems that ensure speedy response to system demands (Figure 3b).

In this scheme the boiler firing receives a megawatt feed forward signal of steam flow or, more commonly, unit load set point, simultaneously with the application of the unit load set point to the turbine loading controller. This strategy is adopted for normal operation when boiler and boiler auxiliaries are in a sufficiently satisfactory state of operation to meet turbine load changes without the need to introduce other features to limit steam pressure increase or decrease. The turbine governor valves are only affected by the steam pressure controller if permissible pressure deviations are exceeded. An additional benefit of the decoupling of the turbine load and pressure control is fewer system parameters to commission and optimise. Where a turbine bypass system exists and provides a pressure limiting function the steam pressure controller is required to provide a pressure support role only.

This form of system has been provided on numerous coal fired stations involving GEC-Alsthom plant. They have the advantage that the unit contributes quickly to frequency support since the limiting influence of the steam pressure on load changing is only active when pre-selected limits are exceeded.

The system accommodates changes in boiler capability. In the event of loss of an major auxiliary (feed pump, PA fan etc.) the system resorts to a supplementary pressure limiter within the governor.

Such configuration has been provided for the 660MW machines at Castle Peak in Hong Kong where with this form of control loading rates of 5% per minute have been achieved in fixed pressure control over a range from 50 to 100% load with pressure controlled to within +/-5 bars and main and reheat steam temperatures controlled within +/-15 degrees centigrade.

Both the previously described co-ordinated approaches use a form of boiler master pressure control in which the turbine demand is a combination of operational setting and frequency generated setting. Grid demand is composed of two distinct components, transient frequency regulation, and long term frequency adjustment. Frequency regulation is the mechanism by which the generator is kept in balance with demand by free governor action so as to limit the magnitude of frequency disturbances. Long term frequency adjustment is a much slower effect which provides means by which generating units participate in generation to maintain system frequency within acceptable limits in an economically scheduled manner. This control is effected either by manual load adjustment or by signals from a remote load dispatch centre.

Since change of electrical output resulting from firing changes is slow due to long boiler inertial time constants, rapid increase in generation can only derive from expenditure of boiler stored energy.

A scheme which permits boiler stored energy to be used for frequency regulation alone has merit for most network situations. This is the one duty for which rapidly available energy is absolutely essential.

As a result of this there has been a move to segregate requirements such that maximum stored energy may be made available to meet system generated transient demands.

This has resulted in the evolution of a scheme known as Active Pressure Governing (Figure 4a). The boiler is loaded by setting the firing demand with the governor valve controlled to maintain the boiler pressure to a set value. The turbine automatically loads when steam is available.

Active pressure governing ensures a maximum availability of stored energy for frequency regulation with consequent benefit to the grid system. It has the advantage that it permits system generated demands for rapid load change and distinguishes clearly between different requirements of this duty and load dispatching. As no other demands are made on stored energy, fluctuations of boiler conditions are reduced.

A scheme provided upon plant which we have recently commissioned involves a unit load set point applied directly to fuel, feedwater and pressure control (Figure 4b). It is applied as a set point to the turbine controls, however, only after being processed through a simple model which represents the natural delayed response between firing application and increased steam generation. This modification to the concept of parallel transmission to both the turbine and boiler controls is that the unit response to load change via load set point control is delayed. This effectively creates a response which is similar in effect to the system of Active Pressure Governing already described. It has a limiting effect on the transient load changing rate of the plant which effectively reduces pressure excursions during loading an deloading and minimises the occasions of introduction of HP bypass control or pressure support action. It also permits for simple introduction of sliding pressure characteristics without the need for complex control terms.

5. STEAM ADMISSION VALVE CONTROL

5.1 Full Arc Admission with Throttle Control

Focus upon improved plant efficiency particularly at part loads has caused attention to be paid to means of reducing inlet loss.

It has been normal practice, particularly in the UK, to provide plant which operate with full arc admission with throttle control and constant pressure operation. Under this mode the flow and, as a consequence, the loads are reduced from valves wide open by equal throttling upon all valves. This is wasteful in efficiency terms since it involves the removal without useful work of energy which has been previously input to the cycle.

Thermal cycling in the inlet region of the machine is more pronounced during this process giving rise to increased impact upon thermal fatigue (or alternatively a need to cycle load at reduced rate to maintain corresponding stress levels).

5.2 Sliding pressure Characteristic

It is more normal today to operate full arc admission turbines with valves wide open in sliding pressure control whenever possible (Figure 5). The TSV pressure has a correspondence with TSV flow. Throttling losses are minimised, resulting in improved efficiency at part loads. In pure sliding pressure operation steam is provided by the boiler at a pressure which is proportional to load and no pressure reduction is necessary across the valves. This provides the benefit of improved efficiency (also tending to make turbine inlet temperatures independent of load) but has the disadvantage that an increase in unit output can only be achieved by increase in boiler firing rate since the steam storage capacity of the boiler cannot be used. Due to the slow thermal response of the complete boiler system it is impractical for plant operating in pure sliding pressure to participate in grid frequency control since the steam system holds no realisable reserves for rapid load pick-up.

5.3 Modified Sliding Pressure Characteristic

To overcome this undesirable operational limitation a modified form of control may be used, essentially, a combination of fixed and sliding pressure operation. In this form of operation the turbine HP governing valves are pre-throttled by a pre-determined amount. This pre-throttling serves to cater for the requirements for sudden load changes. Although this form of control effects a higher pressure load characteristic and consequently a reduced efficiency gain over pure sliding pressure operation at loads below maximum it provides a margin whereby the turbine control system can cause opening of the governing valves as a result of small rapid load increases, rapidly providing a certain amount of the boiler stored steam and enabling the unit to participate in grid frequency control(Figures 5b & 5d)

5.4 Partial Arc Admission

Nozzle control with partial arc admission and fixed TSV pressure has traditionally been used to provide good part load performance at predetermined operating points. Under nozzle control the steam valves open in sequence with a little overlap between them, permitting optimised part load points to be established at which maximum efficiency is created by making use of the lower steam valve throttling drop which occur at these points. This form of control has not been popular in the United Kingdom in the past. It requires specific attention to the machine design for the following reasons:-

- Temperature variation will exist at inlets and nozzle boxes as nozzle groups are introduced.
- Stage 1 moving blades are subject to intense cyclic loading as they pass through active arc regions.
- Dynamic loading of shaft upon bearings necessitates careful attention to sequencing of groups and protection in event of failure of particular governing valves.

This form of steam admission control which has been regularly employed in North America was considered appropriate to a situation where a plant was required to operate for long periods at part load with some reserve capability to meet system strategic reserve requirements. The development of improved management strategies elsewhere in the world which demands that plant be operated efficiently at part load has focused attention upon revised plant design criteria.

Modern machines are specifically designed for nozzle control. Systems which employed nozzle control were mechanically complex in the past. They permitted little flexibility of valve sequencing.

Today the modern governor permits graceful transfer between partial arc and throttle control and vice versa and includes features which transfer control to throttle control and protect the machine in the unlikely event of failure of a control valve causing undesirable inlet steam distribution.

5.5 Hybrid Nozzle Control

Hybrid modes of nozzle control are now sometimes provided where appropriate to the operating regime of the plant. In such a mode a typical configuration could be as depicted in Figure 6 :-

- Fixed pressure nozzle control from 4 VWO at 105% MCR to the 3 arc point at 100% MCR.
- Fixed pressure nozzle control from 3VWO at 100% MCR to the 2 arc point at 90% MCR.
- Sliding pressure operation below the 2 arc point with two governing valves operating in parallel.

5.6 Governor Capability for Independent Control Valve Operations

The flexible nature of the governor control enables valve programmes to be established for manipulation of nozzle control and throttle control. On turbine start-up all four control valves open together at time of introduction of the HP cylinder (at approximately 15% load after initial loading using the IP control valves). When the HP opening demand reaches a threshold value, the No. 3 and 4 HP control valves close under ramp control and the 1 and 2 valves open further to compensate and to maintain the steam flow. The valve opening demand increases until the valves are substantially open and the plant operates in sliding pressure control up to the point where the unit cannot provide additional load in response to demand because maximum pressure has been attained. At this point the demanded valve opening signal increases and the No. 3 valve opens followed by No. 4 valve in the event of a load in excess of MCR being demanded.

As the demand decreases, the sequence remains in 2+1+1 admission until the HP cylinder isolates when the sequence reverts to total admission.

5.7 Performance Benefits

Figure 7 demonstrates the potential efficiency benefits of operating in 2VWO hybrid control compared with the other forms of control which are possible. Figure 8 shows the load pick up capability with the plant operating in 2 VWO Hybrid Operation.

6. COMBINED CYCLE PLANT

6.1 Combined Cycle Operation

The operation of combined cycle plant requires that the steam turbine supplied for such configurations be capable of taking the full steam generated by all the gas turbines in the cycle.

The need for open cycle operation of such plant have upon early installations resulted in the provision of bypass stacks. More recently, however, the capital cost of such features and the perceived unreliability of the transfer control has caused them to be omitted. This imposes additional duty upon the turbine bypass systems required for starting and continued operation of the plant following severe transients and also a significant responsibility upon the control system which must be arranged to control the plant under a range of operating configurations.

The control complexity increases when several gas turbines are included within the cycle, particularly where reheat cycles are included. It is well recognised, that 3 pressure configurations with two or three gas turbines can suffer from inadequate steam generation during cold starting as result of a need to meet minimum steam flow and temperature conditions which are not within the capability of the gas turbine temperature matching control when operating with a single gas turbine.

The problem is basically that sufficient steam is not available at acceptable temperature condition to permit the requirements for introduction of the HP turbine to be met under all types of start. It is normal practice to isolate the HP turbine of machines provided with HP/LP bypass systems during start up to avoid windage heating within the turbine blading.

In common with conventional fossil fired reheat plant provided with HP/LP bypass systems steam turbines included in combined cycles can suffer loss of the natural pressure ratio across the HP turbine due to the operation of the bypass causing artificially high reheat/HP exhaust pressures.

When all the steam which is available cannot pass through the turbine cylinders the surplus is routed to the condenser through the bypasses. Under these conditions the reheat steam pressure is controlled by the bypass system. The operating pressure of the bypass affects the cold reheat pressure and hence the HP exhaust pressure.

Operation at low reheat pressure provides for a greater possible pressure differential across the HP turbine and promotes more secure operation. For safe operation of an HP cylinder (or any other cylinder for that matter) it is necessary to ensure a minimum steam flow. If this cannot be assured it is necessary to evacuate it to prevent windage heating.

To protecting our turbines against HP windage heating we provide means of isolating the HP turbine whilst running to speed and at low loads and provide substantial venting between HP cylinder and condenser with means to control it remotely.

We provide control within the governor which balances steam flows such that when the potential demanded HP steam flow required to meet the current load demand exceeds the minimum allowable flow as a function of reheat pressure and is less than that flow being supplied by the HP Bypass System the suppression of HP governing valve opening is removed allowing the HP to load up.

When the HP turbine is introduced it is necessary that steam to metal temperature criteria are met. High steam chest differential temperatures can occur as a consequence of the necessary gas turbine loads required to achieve target steam quantities. If measures are not taken it is likely that sufficient steam will not be available at acceptable temperature condition to permit the requirements for introduction of the HP turbine to be met. Attempts to operate under such marginal conditions can be the cause of windage heating during HP turbine introduction and drop off in main steam pressure causing pressure support functions to be activated.

The situation may be improved by:-

a) Provision of more steam during starting.

   Steam availability is improved by:-

i) Addressing Gas Turbine operation to maximise steam availability at lower steam temperatures.

ii) Increasing allowable steam temperatures for cold and some warm starts by use of a period of forward warming of the HP turbine lower speeds where windage heating is not a problem.

b) Ensuring effective utilisation of steam.

Full utilisation of all steam available is made by:-

ii) Ensuring that the HP turbine takes all available steam immediately following HP turbine introduction by loading the turbine to a point where HP valves are forced into pressure control on introduction of the HP turbine.

iii)Ensuring that the IP turbine passes all steam available on introduction of the HP turbine by causing the IP valves to become forced into pressure control. This does two things - it maximises the flow through the IP turbine and due to the deadband between IP bypass valve set pressure and IP governing valve pressure support set pressure ensures the minimum reheat pressure.

c) Minimising load at which HP turbine may be released:-
Provision of IP Bypass Valves of sufficient size to allow the minimum reheat pressure which is acceptable to the boiler supplier. This will permit HP turbine introduction to be made at a lower steam turbine load/flow.

6.2 Single Shaft Combined Cycle Units

The control of the single shaft combined cycle unit presents a number of unique control problems when compared to other forms of steam turbine control.

The governing functions associated with the steam turbine may be regarded as much simpler than these of the conventional machine. Its function, once the unit has been synchronised to the grid having been run up using a mix of static frequency converter (SFC) and gas turbine, is to ramp open the steam valves and allow the turbine to load, initially in throttle control and finally in sliding pressure control until full load is achieved. It has no involvement in overall block control. Unit load control is achieved using a master controller located within the DCS (Figure 9). The Unit Master Controller loop adjusts the unit total load set point given by the operator from the Central Control Room. This control loop compares the unit load set point coming from the unit master controller with the actual load of the combined cycle and sends load change commands to the gas turbine fuel controller.

The ST follows the GT load variations with a time delay depending mainly on the HRSG inertia.

Therefore initial load change is accommodated by the gas turbine, then as the ST load changes the booster unit master controller pays off the GT output to maintain block output constant.

In normal operating conditions, the combined cycle can operate in two modes:

- load control: In this case, the power delivered is equal to the set point value; after a grid frequency variation in which the GT has participated, the unit master load controller redresses the power to the set point (± 0.5 MW).

- frequency participation mode: after a grid frequency variation in which the GT has participated following its droop factor, the combined cycle load set point is modified according to the frequency difference.

During start-up of the unit, the Unit Master load controller adjusts the exhaust gas temperature in order to provide acceptable steam temperatures. This exhaust gas temperature set point depends on the ST thermal state.

The overall control system:

- Provides automatic loading of the Combined Cycle plant during start-up from shutdown or other steady state to normal on load operation,

- Provides automatic deloading of the Combined Cycle plant during shutdown.

- Provides a means of controlling the total load of the Combined Cycle plant so that the GT and ST operate as a single block.

- Provides secondary response control during frequency sensitive operation.

7. FREQUENCY CONTROL

7.1 Response to System Disturbances

The application of the modern governor has allowed the adoption of features which improve the response of the plant to system disturbances. This is particularly necessary in developing areas where plant size has progressed at a faster rate than system development. Units of large power rating contribute more to grid support than of lower rating older units. This can lead to the use of large units as base load producers restricting them to constant output operation by use of large governor droop with or intentional deadband.

This continues to be the case with nuclear units, although there is an increasing introduction of reactor thermal power automatic control systems for grid support (for example the Grey Rod Control of certain PWRs which combine the turbine and reactor control systems to provide limited range unit manoeuvrability to change in frequency).

This is not an acceptable situation for rapidly developing utilities. The use of programmable control systems with high performance fast acting high pressure steam valve actuation units have allowed systems to be implemented which more closely match the generator response to grid needs thus reducing the risk of discontinued operation as result of severe disturbances and helps assure the continuity of electrical supply to the grid.

Since its early form the electronic governor has unquestionably brought benefit to steady state stability of the unit operating within the network. and the advent of state of the art digital excitation systems have brought further improvement. Most AVRs provided on new plant today includes a power system stabilisation function as a matter of course. It is, however, to improvement in ability to cope with transient stability that the governor designer has turned in recent times. Such transients result from system fault load swings and tripping of sections of the network.

Requirements for instantaneous and sustained fast valving demand that the valve actuation systems may reposition steam valves accurately and rapidly to new load settings in response to demands of remote switching actions or the governor's own internal frequency monitoring. In such cases the response of the unit has been improved by eliminating "on/off" accelerometric terms, replacing them with continuous dynamic corrective actions.

This is only possible if the performance of the speed control system and servomotors is of a sufficiently high level. The servomotors must be capable of partial and rapid closure unlike those equipped with an additional quick dump valve which always cause full closure of the steam valves. The GEC Alsthom servomotor is pilot controlled by an electronic position controller actuating a large flow servovalve as illustrated in Figure 10. The figure also shows typical responses of a servomotor equipped with such a servovalve.

7.2 Combined Cycle Frequency Control

Due to the nature of the plant and operating regime the steam turbine governor role in frequency response is very limited.

In normal operation:-

- There is no external limiting factors (e.g. from HRSG) on GT response to frequency changes. The GT exhaust profile is relatively flat betweem 30% and 100% load.

- The steam turbine is in sliding pressure operation. Therefore during transients there is no initial change in ST output due to thermal inertia in HRSG.

Primary response is achieved by using a lower GT governor droop setting. (Typically 2.7% - equivalent to a block droop of 4%).

Secondary response is achieved using a DCS to control block output, droop setting of 4%. The DCS controls block output by adjusting GT output to compensate for changes in ST output.

The steam turbine governor operates with a droop setting of 6% but has a +/- 0.5 Hz deadband effectively limiting steam turbine frequency response.

7.3 NGC Requirements

The requirement for Frequency Response Capability by NGC imposes requirements upon the Generators which have presented some problems for plant which employs gas turbines.

The requirement for provision of decrease in output pro rata with frequency falls below 49.5 Hz is readily met by steam turbines whose natural characteristic more than adequately meets the NGC requirement. The gas turbine being an air breathing machine whose mass flow is speed related suffers from the fact that with a limit on firing temperature the resultant load drop off is governed by a N^2 relationship, and does not naturally meet the requirement.

Meeting this requirement is difficult for all manufacturers. The GT load control alters the gas mass flow to maintain turbine inlet temperatures at set point to maintain optimum efficiency. However if temperature limits are exceeded the temperature control interprets a firing limit. Without application of controlled overfiring the requirement cannot be met and the plant is non-compliant. If the machine is incapable of accepting such overfiring the requirement can only be met by derating the plant. Figure 11 shows the natural characteristic of a gas turbine to frequency and the application of controlled overfiring to achieve compliance. Overfiring will have only minimal impact upon plant life since life erosion only occurs when actual exhaust temperature exceeds base rated temperature.

For some manufacturers gaining compliance is achieved by switching to peak operation if the system frequency falls below a small setable level. This is a compromise between likelihood or frequency of the event occurring at elevated ambient temperatures and module life reduction.

8. CONCLUSIONS

I believe that we have shown that the development of governing control during the past 30 years has anticipated and met the requirements of the industry. This has been due in, no small part, to the early introduction of electronic and microprocessor control. The nature of this evolution has resulted in the electronic governors of large steam turbines fulfilling a variety of tasks which might otherwise have been invested in other automation equipment. Whilst a need remains for the industry to introduce greater levels of automation to enable reduced manning levels, higher efficiencies and better co-ordination of the plant it is almost certain that more control will be invested within DCS equipment in the future, leaving the steam turbine governor to look after the fundamental control of frequency, load and steam valve position.

Figure 1

STATION DCS
MODBUS SERIAL LINK
UPPER LEVEL
SPEED SET POINT
LOAD REFERENCE
HP & IP VALVE OPENING LIMITS
MANUAL CONTROL
+
-
MANUAL CONTROL
+
-
MANUAL CONTROL
+
-
AUTO
MAN
AUTO
MAN
AUTO
MAN
ACTUAL SPEED
N1
SPEED CONTROL & VALVE OPENING SIGNAL PROCESSING CHANNEL 1
N2
SPEED CONTROL & VALVE OPENING SIGNAL PROCESSING CHANNEL 2
N3
SPEED CONTROL & VALVE OPENING SIGNAL PROCESSING CHANNEL 3
HP VALVE OPENING CONTROL SIGNALS
IP VALVE OPENING CONTROL SIGNALS
2/3 VOTING DEVICES
2/3 VOTING DEVICES
HP VALVE POSITION CONTROLLERS
IP VALVE POSITION CONTROLLERS
HP GOVERNING VALVE SERVOMOTORS
IP GOVERNING VALVE SERVOMOTORS

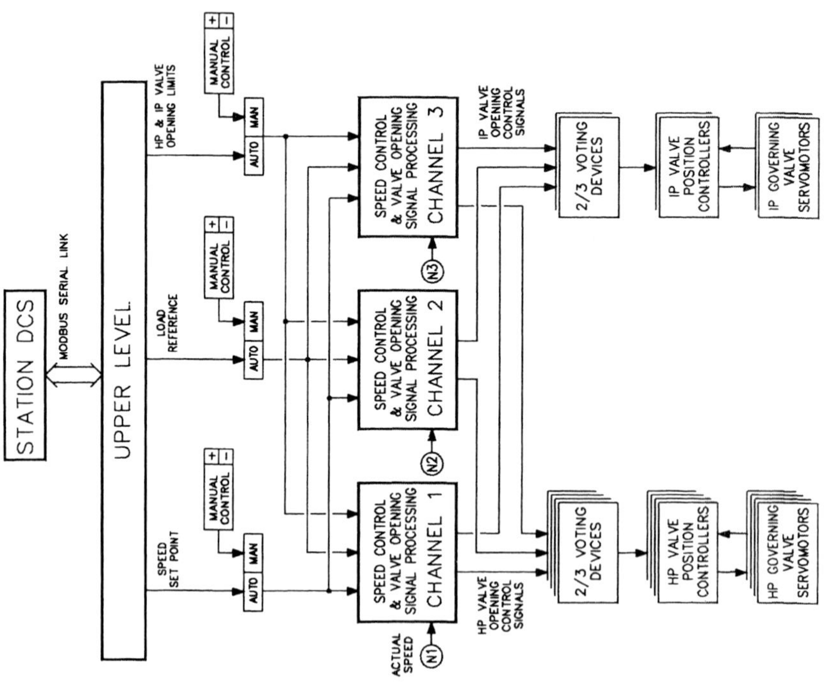

Figure 2

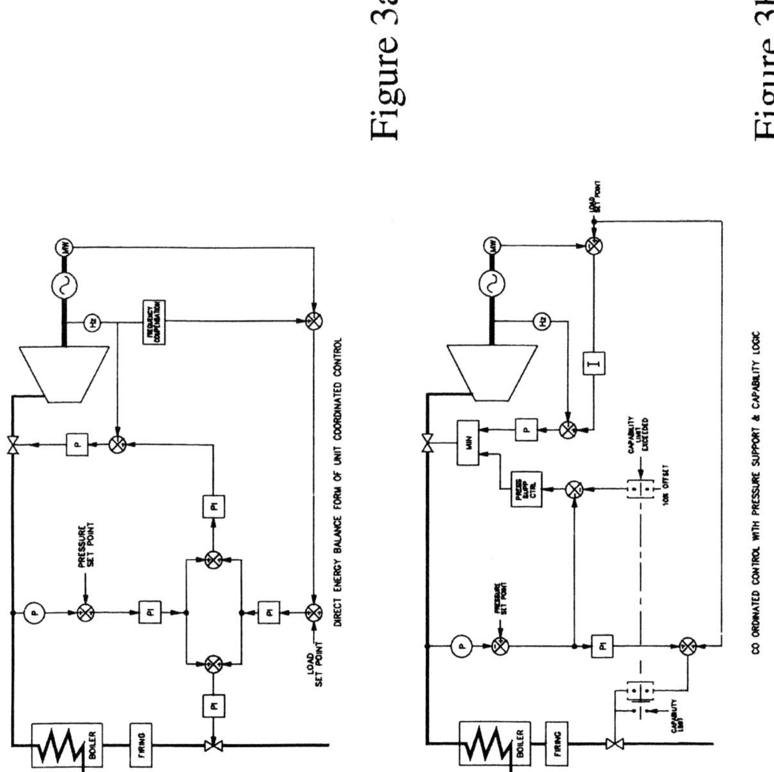

DIRECT ENERGY BALANCE FORM OF UNIT COORDINATED CONTROL

Figure 3a

CO ORDINATED CONTROL WITH PRESSURE SUPPORT & CAPABILITY LOGIC

Figure 3b

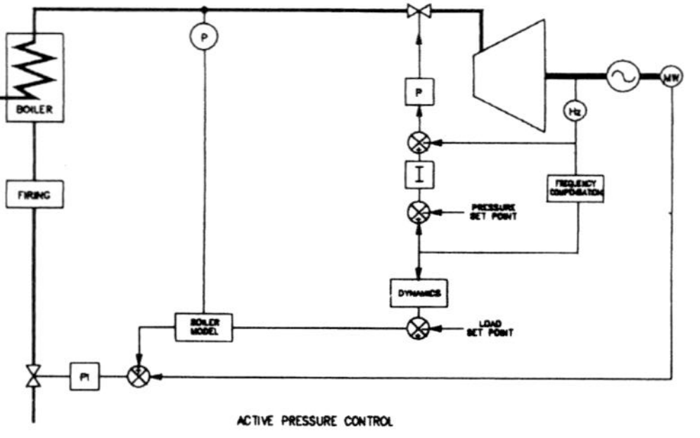

Figure 4a

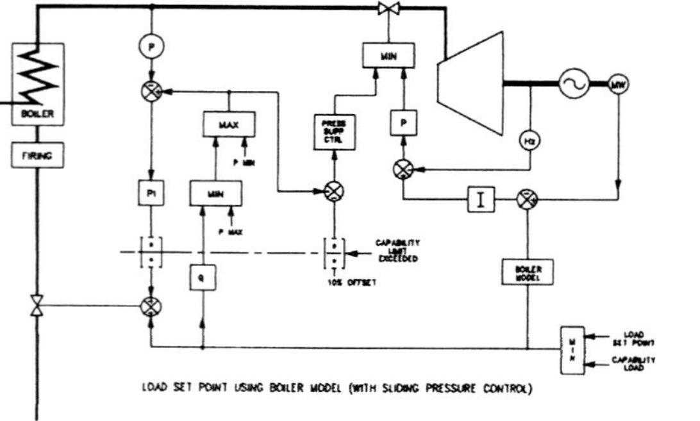

Figure 4b

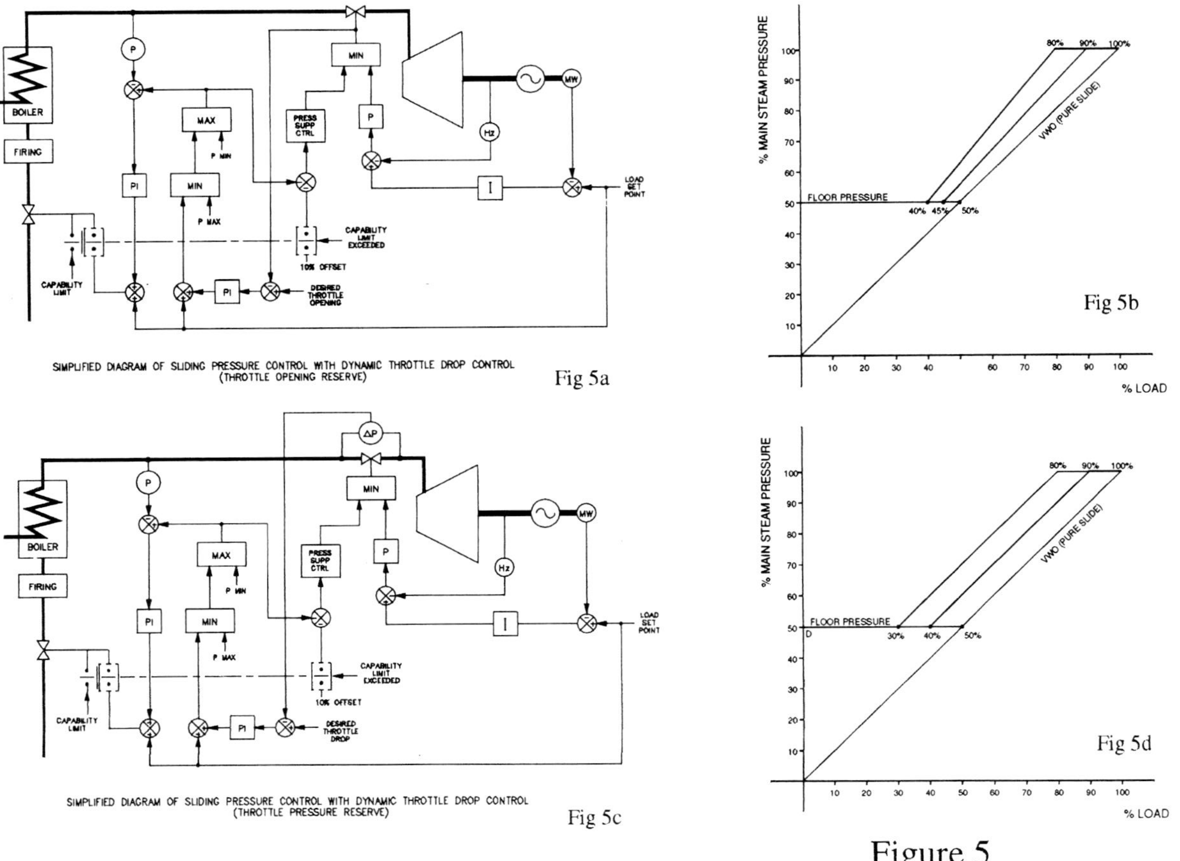

SIMPLIFIED DIAGRAM OF SLIDING PRESSURE CONTROL WITH DYNAMIC THROTTLE DROP CONTROL (THROTTLE OPENING RESERVE)

Fig 5a

Fig 5b

SIMPLIFIED DIAGRAM OF SLIDING PRESSURE CONTROL WITH DYNAMIC THROTTLE DROP CONTROL (THROTTLE PRESSURE RESERVE)

Fig 5c

Fig 5d

Figure 5

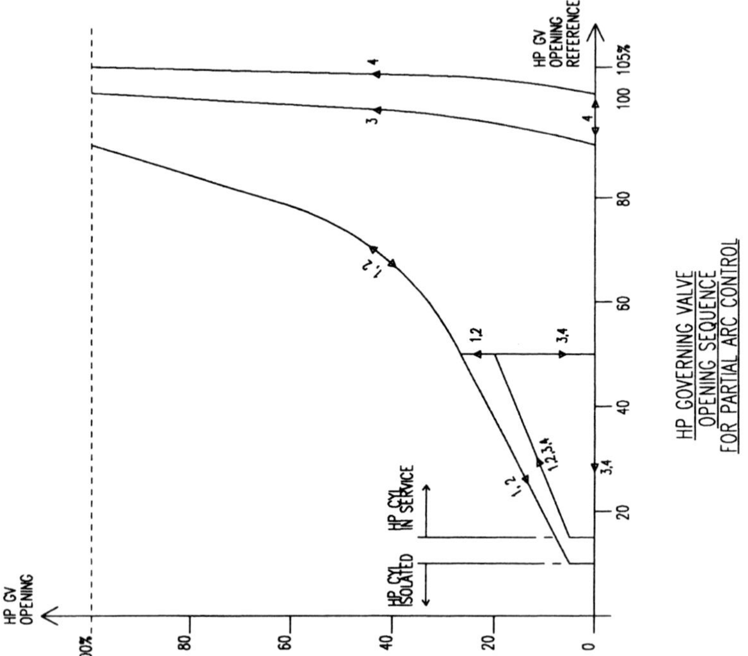

HP GOVERNING VALVE
OPENING SEQUENCE
FOR PARTIAL ARC CONTROL

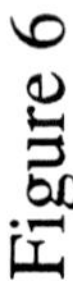

Figure 6

HEAT RATE V LOAD FOR DIFFERENT CONTROL MODES

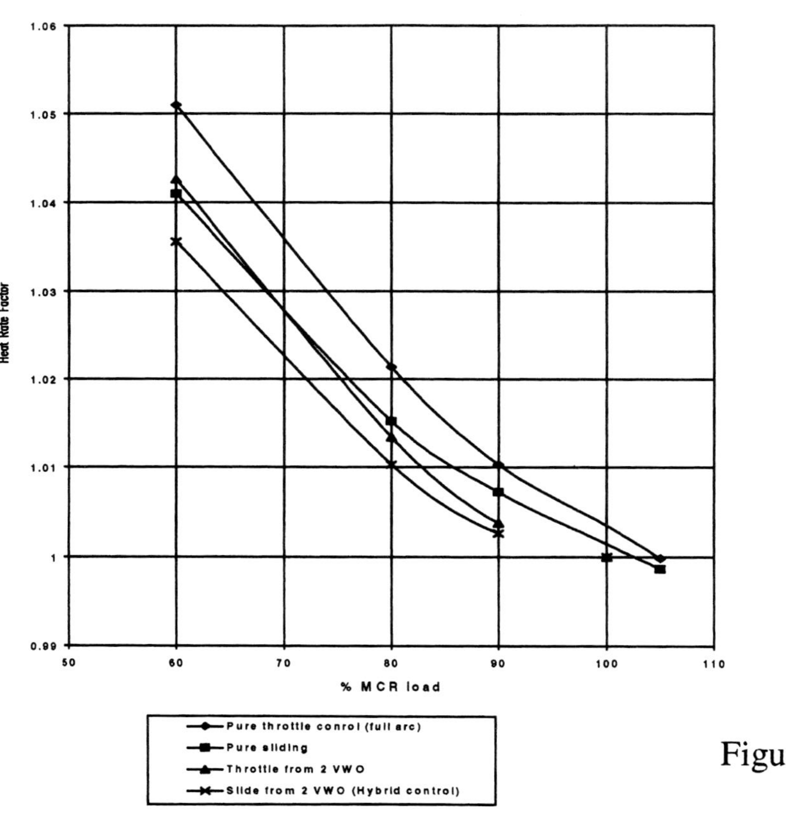

Figure 7

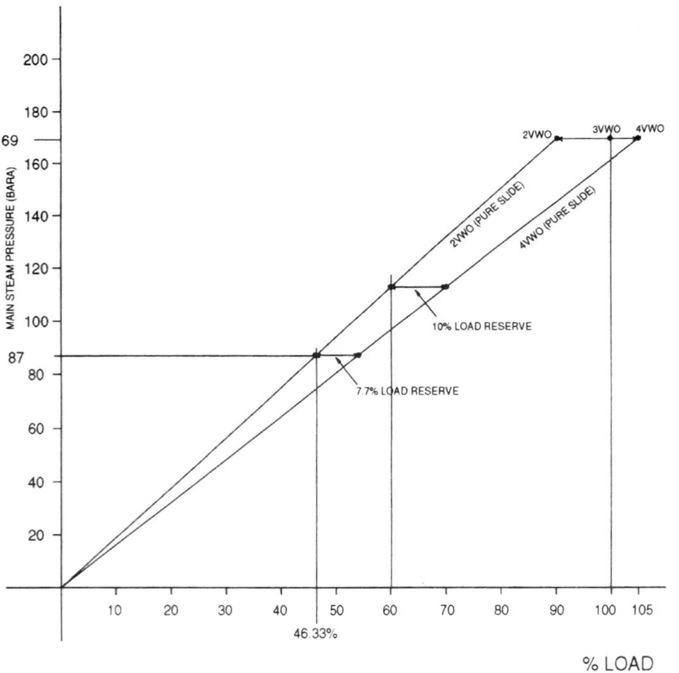

SLIDING PRESSURE CHARACTERISTIC
HYBRID OPERATION WITH PARTIAL ARC ADMISSION

Figure 8

LP
HP
IP
GT
MW_A
P
Tm
Hz
TE
TA
MW_{SP}
f(x)
f(x)
f(x)
f(x)
X
3000 rpm
25
DROOP CONTROL
ON/OFF FREQUENCY PARTICIPATION
LOAD CONTROL MODE
EXHAUST GAS TRIM CONTROL MODE
&
&
OR
GAS TURBINE FUEL CONTROLLER
Tm Stean Turbine Metal Temperature
TE Gas Turbine Exhaust Temperature
TA Ambient Air Temperature
MW_A Actual Load
MW_{SP} Desired Load
UNIT MASTER LOAD CONTROL

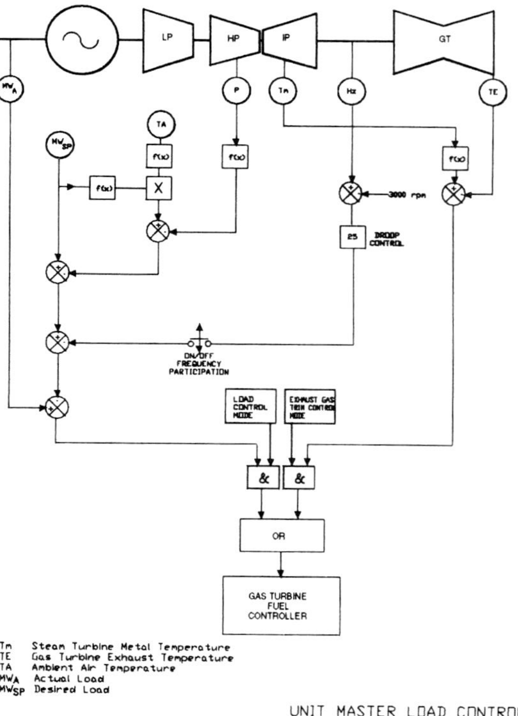

Figure 9

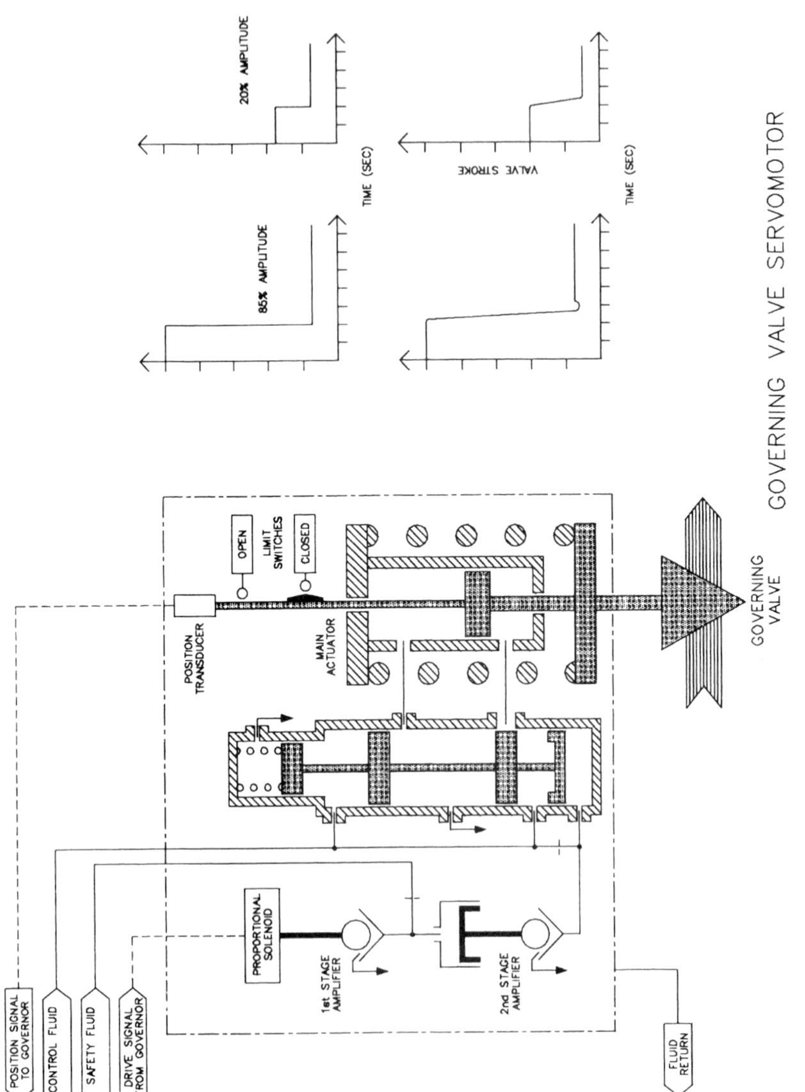

GOVERNING VALVE SERVOMOTOR

Figure 10

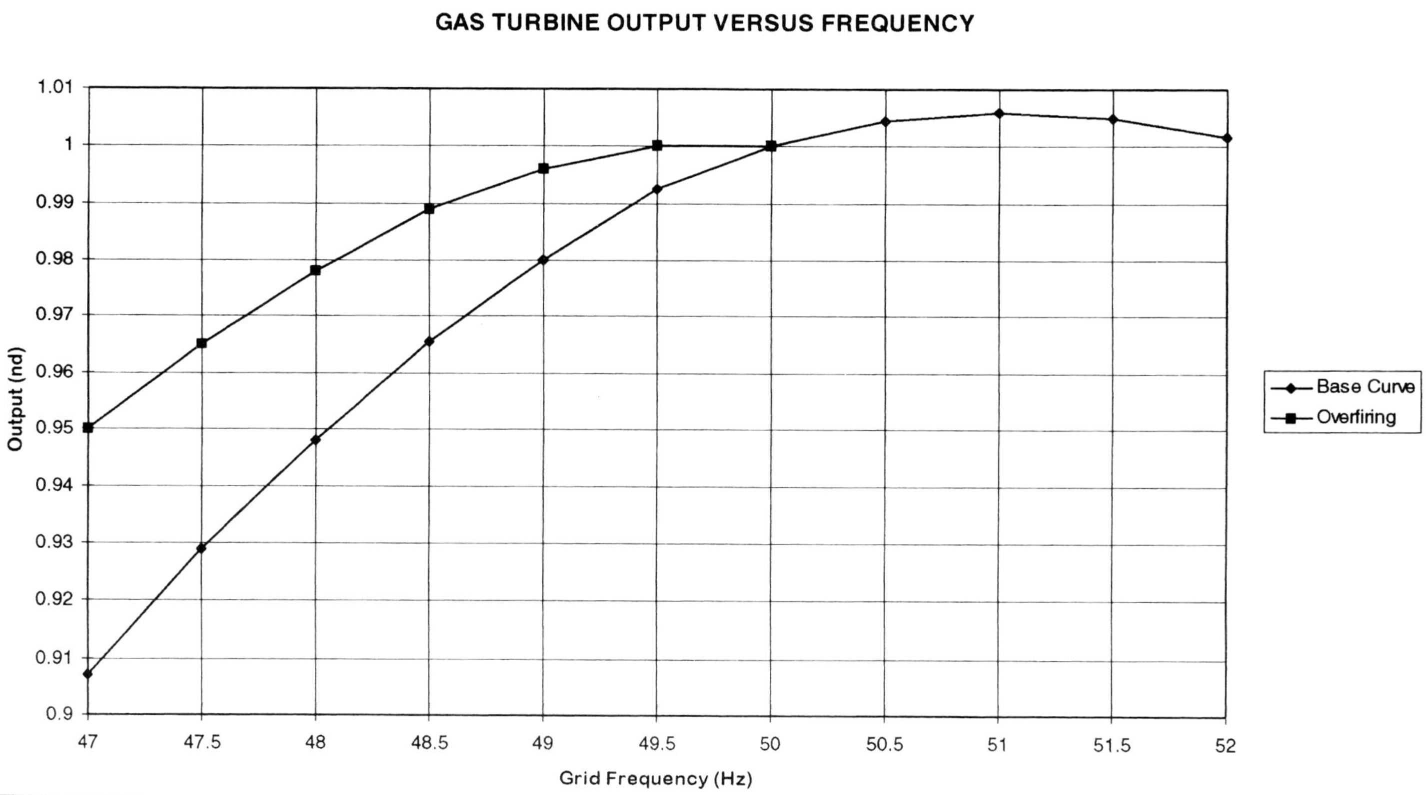
GAS TURBINE OUTPUT VERSUS FREQUENCY
Output (nd)
1.01
1
0.99
0.98
0.97
0.96
0.95
0.94
0.93
0.92
0.91
0.9
47
47.5
48
48.5
49
49.5
50
50.5
51
51.5
52
Grid Frequency (Hz)
Base Curve
Overfiring

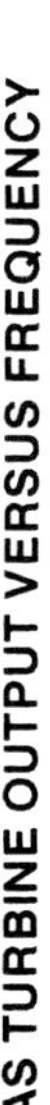

Figure 11

S548/003/98

A state-of-the-art triplex modular redundant controller for electro-hydraulic governors and excitation systems

N J GREEN and **J M OLDS**
Rolls-Royce Industrial Controls Limited, Gateshead, UK

1. INTRODUCTION

Rolls Royce Industrial Controls Ltd. have been supplying a complete range of turbine island products for in excess of 25 years. Of particular importance in that range are electro hydraulic governors (EHGs) and excitation systems (AVRs).

The early systems were based on analogue controllers but have been exclusively digitally based over the last 10 years.

In response to customer requirements these systems have been refined over the years to:-

- reduce costs.
- simplify the design procedures.
- and to achieve higher availability.

This paper reviews two possible redundancy architectures and outlines how the equipment has developed over the last few years.

2. POSSIBLE REDUNDANT ARCHITECTURES

Redundant architectures are designed to maintain all essential system capability with a single component failure, and to degrade in some defined way if a subsequent fault occurs before the first fault has been rectified.

The two redundancy schemes most commonly applied to EHGs and AVRs are:-

1. Main/Standby Redundancy
2. Triplex Modular Redundancy

These schemes are now discussed in more detail, with particular emphasis on methods of operation, fault detection and fault reporting.

2.1 Main/Standby Redundancy

This system consists of two identical control channels with only one supplying the output at any time. This scheme is perhaps the most commonly occurring redundancy arrangement. The scheme's Achilles heel is it's fault detection and the subsequent main to standby transfer.

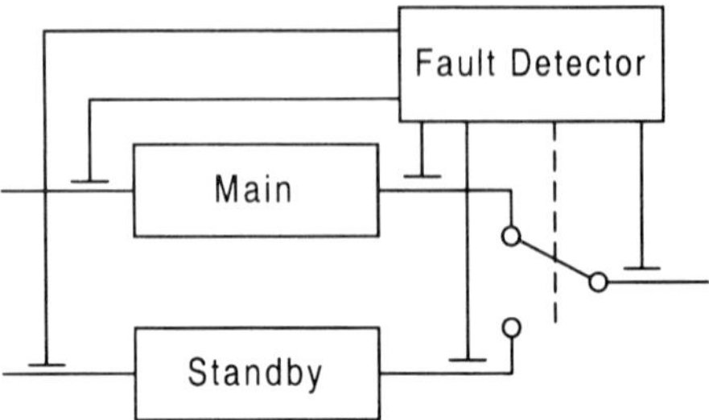

The fault detector needs to be able to identify that the main channel is not controlling correctly. To do this it requires information on the inputs to the "main", output from the system, and also needs some knowledge of the control algorithm and/or the characteristics of the plant. With any two channel system there are a proportion of faults which can only be identified by introducing an independent (third) measurement channel and comparing the third measurement with those of the two control channels.

Fault reporting is required for both the main and standby systems. In the case of the standby, adequate fault reporting is often very difficult to achieve. If the main/standby system is symmetrical it is possible to regularly carry out a symmetrical swap of the main/standby functions and this will then reveal any faults which the fault reporting system was unable to discover.

For complex systems the extent and reliability of an adequate fault detector can become a matter of some concern, and it is reasonable to say that main/standby schemes are most attractive for systems which do not have complex fault detection requirements; i.e. have single input, simple control algorithm and low complexity plant.

2.2 Triplex Modular Redundancy (TMR)

Three identical control channels with the median (e.g. middle) channel selected as the output; the median selector effectively isolates single faults from the output. The fault detector is in effect replaced by a median selector which does not require any knowledge of the system inputs, control process or plant.

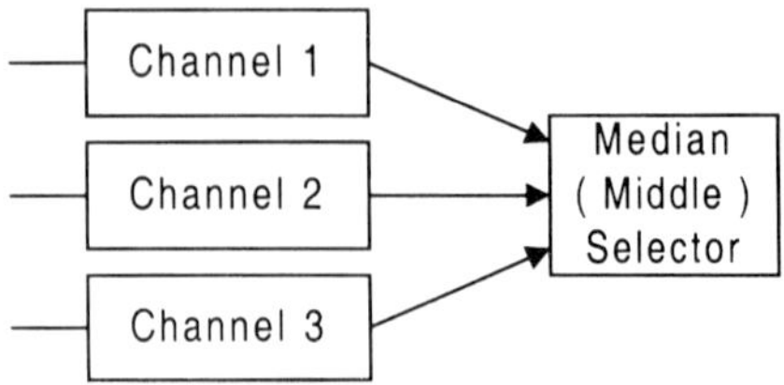

Under normal circumstances in a TMR system any one of the three channels could be supplying the output at any instant in time, and the selected channel will randomly change as a result of small differences between channels and/or noise components. In effect all channels contribute to the output and their ability to control the system is exercised on a regular basis. This feature of TMR systems is not widely understood by engineers.

Fault reporting is still necessary, but this is simple and unambiguous using channel output cross comparison techniques. Although such fault reporting will identify a channel fault, it will not highlight the cause of the fault within the channel and additional fault reporting will be required to facilitate maintenance.

Once a faulty channel has been identified an alternative strategy for the use and detection of faults in the remaining two channels must be adopted. The most obvious approach is to use the average of the two remaining channels while monitoring that the difference between them is within some permitted band. If the difference exceeds the band the system can no longer take any logical control action and the system should be tripped.

If after a single channel fault has been identified, that channel returns to within a acceptable tolerance band, then it is possible to automatically return to full median selection; although the fact that a fault had occurred should be reported.

TMR schemes are able to handle multi-input complex control systems with relative ease with the additional benefit that they tend to "fail safe".

2.3 Availability Comparison

When considering the availability of redundant systems it is important to use a method which correctly takes into account repair times. One such method is that proposed by Markov in 1907 which divides the system into a number of "states" and then considers the "rates of transition" between the states to build up a simulation of the system behaviour.

The simulation involves a parameter known as "coverage". This is defined as the probability of re-configuring a system successfully when a fault occurs. Coverage is a real measure of how well the system designer implemented fault tolerance, and is difficult to accurately measure without a lot of detailed information about the internal system architecture. A perfect system that can withstand all faults will have 100% coverage. In reality this is not possible and figures of 99% and 97% are typical published figures for triplex and duplex architectures respectively.

The following table shows the MTBFO comparison for the different redundancy schemes assuming that the MTBF of one channel is 5 years.

Redundancy		Triplex	Main/Standby
Coverage		99%	97%
Repair time (hours)	Reboot time (hours)	MTBFO (years)	
2	6	27.5	13.9
4	8	20.5	10.4
12	16	9.9	5.2

The tabulated figures clearly indicate the superior fault tolerance of triplex redundancy when compared to other arrangements.

2.4 The Chosen Redundancy

As described above a single median selector is inferred and this has the potential for single point failure. In reality it is almost always possible to implement the system with several median selectors working in parallel and this eliminates the potential for single point failures. The EHG and AVR architecture described later in the paper all use parallel redundancy in the output stages and at least some of the fault reporting is assigned to these output stages to ensure fault reporting under most multiple fault situations.

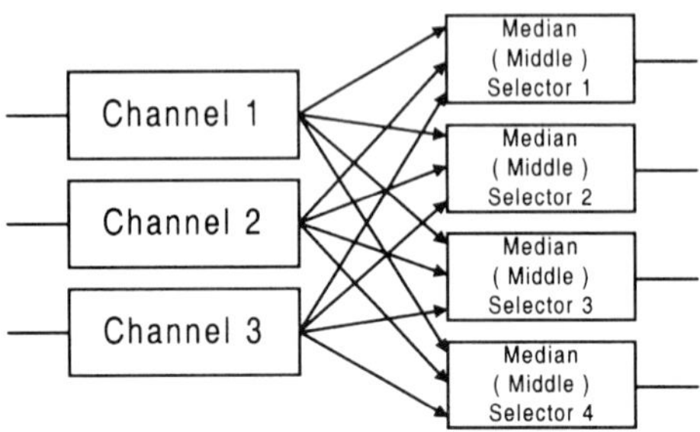

Triplex systems are generally perceived to be more expensive than the alternatives, but in reality when the fault detection associated with the other arrangements is taken into account the difference is small and in general the advantages of a TMR system fully justify the extra costs.

For control systems of this type Rolls Royce consider that triplex modular redundancy is superior in all respects and as a consequence when high availability is a customer requirement will only support TMR designs for both EHGs and AVRs.

Where cost is the over-riding criteria single channel designs are available but even then attempts are made to introduce some form of redundancy.

3. HARDWARE

3.1 Hardware & Obsolescence

Obsolescence of electronic components is a problem that designers and operators need to consider carefully. The technology used in modern electronic systems is advancing so quickly that obsolescence will be met after typically 5 to 8 years, and even earlier with some of the first generation digital controllers.

In principal there are two techniques possible to counteract obsolescence:-

- it is possible to use industry standard electronic modules which in theory would be supported by functionally equivalent modules for the foreseeable future.
- there will be some electronic modules that are specifically designed for the system and spares holdings should then relate to the predicted life of the equipment.

In general terms, the magnitude of the obsolescence problem is proportional to the number of different types of modules that are used in the particular implementation.

3.2 The Previous Generation Hardware

The previous generation equipment is typical of systems constructed from a number of modules where each module carries out a specific electronic function. In the Rolls Royce case the redundancy issues were addressed by using industry standard VME modules, and 5 types of modules plus a valve controller were employed. In addition power supplies are required but these are not shown as they are not normally of concern with regard to obsolescence.

3.3 Latest Generation Hardware

As stated the magnitude of the obsolescence problem is proportional to the number of different types of modules that are used. It is clear that reducing the number of module types used in the system is of prime importance.

By using a state of the art micro-processor (Texas DSP) and using a dedicated mother board it has been possible to design a single module that has **ALL** input/output capability and power to act as one channel of an EHG or AVR, or in fact both an EHG & AVR at the same time. A triplex design can then be constructed from three of these modules.

The typical EHG system shown below uses one type of modules plus a valve controller. Power supplies are excluded as before. The obsolescence problem is obviously much easier to overcome by stocking spares modules relative to the life of the equipment.

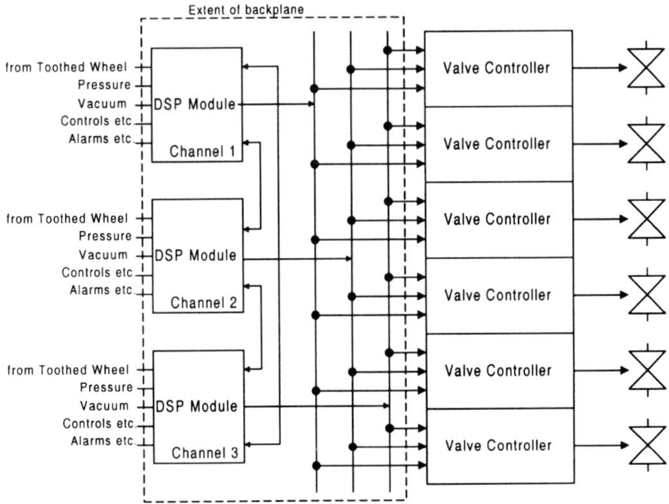

Typical EHG Implementation

With only one module per TMR channel there are considerably less interconnections required to implement the system and the opportunity can be taken to incorporate all the inter connections within a printed circuit board back plane that additionally includes the plant terminals. A wire count on the previous equipment indicated 1700 wires, a similar count on the latest equipment indicates less than 50 wires. Implementing the interconnections on a PCB back plane drastically reduces the assembly times and also eliminates the possibility of wiring faults, and associated testing and debugging times are proportionately reduced.

With a reduction in the number of modules and a reduction in inter-connections there is a corresponding increase in reliability and availability. The 'forced outage rate' will as a result be improved considerably.

4. SOFTWARE

4.1 Previous Generation Software

Techniques for generating good structured software have progressed in leaps and bounds over the last ten years but never the less a typical EHG may typically contain 50 different time shared processes and each process could involve 2 or 3 source files. A software project of this size needs to be carefully controlled and specialist programmers need a detailed knowledge of the system configuration in order to carry out modifications and maintenance.

The dynamics of the controlled system dictate that software update times of the order of 10 mSec are required for EHGs and AVRs and the software run times must be considerably less than this. Normal techniques used for writing software may be too slow to achieve the required run times and alternative techniques which use less time must be developed and adopted by programmers.

In brief the problems in controlling such projects, maintaining software quality and speed, and retaining the required levels of specialist staff are extremely onerous.

4.2 Latest Generation Software

The latest equipment relies heavily on system modelling techniques and auto code generation to simplify the task of programmers. Models of the controller and the controlled system are constructed, tuned and de-bugged before extracting the controller model and generating code automatically.

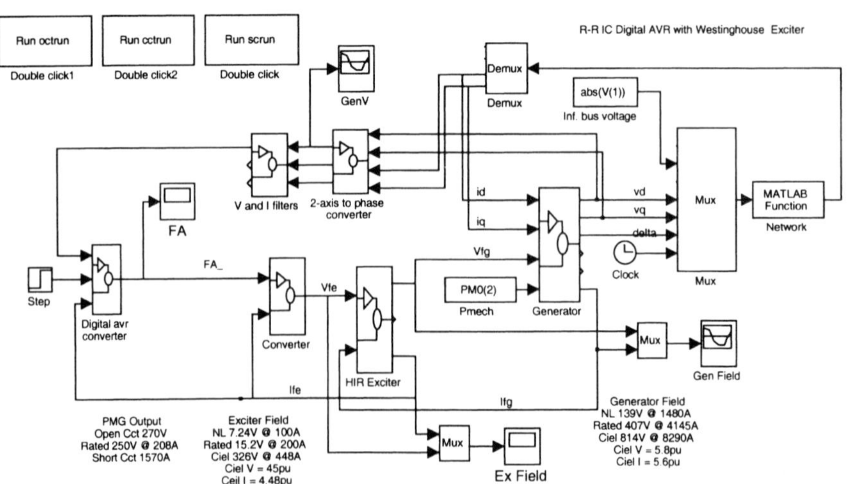

Typical Model of AVR, Thyristor Converter, Exciter, Generator & Grid System

Input and output software is generated manually but is fixed and does not change for different applications. As a consequence it is possible to exhaustively test this relatively small amount of software.

Significant gains are possible:-

- Structural or functional errors are identified & eliminated at the modelling stage.
- The final code is effectively free of coding errors.
- The design documentation is generated automatically e.g. block diagrams.
- Testing time is proportionately reduced.
- Specialised software engineers are not required and engineering resource can concentrate on the application of the controller.
- The selected micro processor is fast enough to eliminate concerns over run times.

The DSP micro processor selected for these applications is very fast (estimated at 40 times faster than the previous generation) and as a result special techniques to ensure fast run times are not necessary. In reality it is possible to update the AVR output every 3.33 mSec to ensures the fastest possible response. However 10 mSec update for the EHG is considered adequate.

It has been estimated that the previous design involved 3 man years of effort and the equivalent figure for the new design is closer to 3 man months. This saving in man power is reflected in the selling price.

5. APPLICATION TO AVRS

5.1 Redundancy in the converter

It is often difficult to see how a TMR controller can be successfully mated with a 3 phase thyristor converter without compromising the concept of 'no single point of failure'.

The most common technique is to employ two 100% rated thyristor bridges running in parallel. In order to maintain the equipment it is necessary to supply two 5 pole isolators. In total the equipment supplier has to provide 200% capability and expensive 5 pole isolators in order to meet the redundancy criteria of 'no single point of failure' and the ability to maintain on line.

If the control system produces firing pulses then it is a simple task to locate a 'median firing pulse selector' at each thyristor and to mount each thyristor in a plug in module. With this arrangement the redundancy criteria can typically be met with a total of 133% capability and without the need for expensive 5 pole isolators.

Typically each thyristor arm (shown as a thyristor in the drawing over) in fact consists of several thyristors and median selectors in parallel and a typical large static excitation system could contain 24 thyristors/median selector modules. This system approaches the ideal redundancy system at realistic cost.

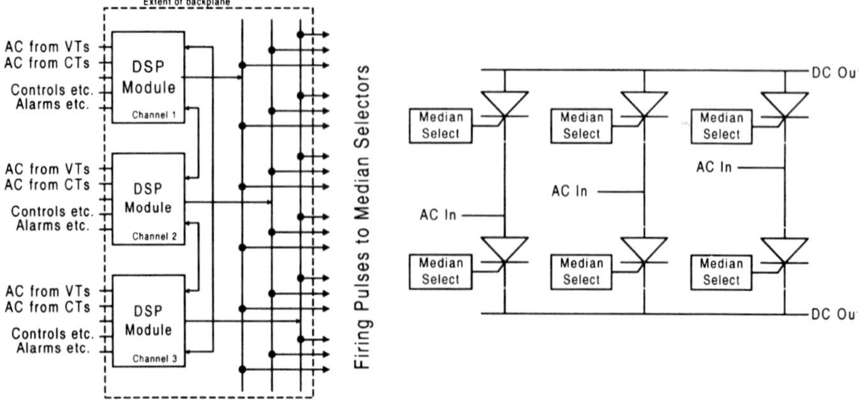

6. APPLICATION TO EHGS

6.1 Introduction

The electro hydraulic interface to the turbine is as important to the MTBFO of the unit as the redundant control channels.

As has been stated above, the final output median selection and channel fault detection in an R-R TMR EHG system is performed by each valve controller. One controller is usually allocated to each steam control valve.

Typically, a large steam turbine may have four parallel steam inlet paths. Each inlet path will typically have a control valve prior to the HP and IP stages, and the unit may be run to speed under control of the HP safety valves.

Retrofit EHG's generally fall into one of two categories, 'Conversion' or 'Replacement'. A summary case study of a typical 'Conversion' scheme is included towards the end of the paper.

6.2 Valve Control Configurations

There are many factors which have influenced the steam valve interface configuration for large steam turbines over the last thirty years. These include:-

- Size (number/configuration of cylinders)
- Available technology and materials
- Accuracy and actuating forces
- Availability (parallel inlet paths)
- Efficiency (partial arc operation)
- Grid response/linearity (flyball-electronic)
- Interfacing (modulating control, automated testing)

There are many and varied configurations of valve control schemes in operation and it is not unusual to find two similarly rated sets from a single manufacturer with wildly differing control arrangements.

6.3 Features of a 'Standard' Valve Controller design

A 'base design' of valve controller module is therefore required with sufficient adaptability to be configured to suit all potential application. This 'base' or 'standard' controller attributes will have the following features:-

Control

- Triplex or simplex inputs (TMR or Simplex)
- 0.1% control resolution, better than 10mS loop cycle time (fast, accurate control)
- Ease of servo drive booster interfacing (larger servo designs)
- Single or dual loop position feedback (voltage/current)
- P or PI control (minimise errors in single loop control)
- Duplex control interface flexibility (for critical simplex EHG servo applications)
- Typical pilot valve slew time 5mS, main ram = 100mS. (closed – dual loop control)
- Manual control 'back-up' valve stroking (useful as standby mode for simplex)
- Servo loop compensation (overcomes servo resonance)
- Selectable dither (frequency/amplitude)
- Selectable non-linear gain (to suit control port design)

Fault Detection

- Configurable fault detection with transient capture (to suit application)
- Maintenance port for on-line condition monitoring/fault diagnosis ('soft' faults)
- Communicate status and fault information (operator status info on EHG)

Hardware

- Cost effective and available hardware.
- Work in multi drop mode (serial address and data)
- Have OLVT capability (on line stroke testing and re-instatement)
- On board simulation circuitry (minimises test interconnections)

6.4 Existing Controller - Overview Block Diagram

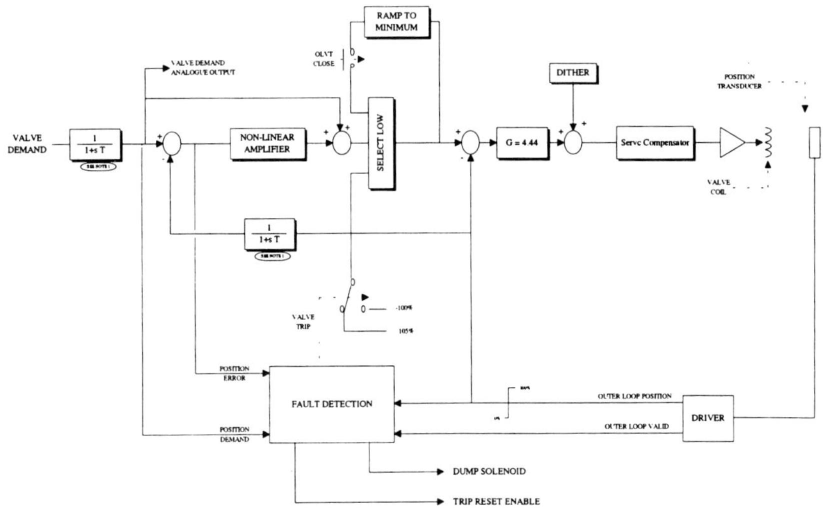

<u>**STEAM VALVE CONTROL MODULE (single control loop)**</u>

6.5 The next generation of valve controller

The principles of using system modelling to identify/optimise the control strategy leading to automatic code generation onto the target hardware platform can equally be applied to the valve controller. For the valve controller the two major areas of interest are:-

- Control Algorithm
- Fault Detection

6.6 Improvements to the existing control algorithm

The existing controller design is summarised above. The fundamental limiting factors in the existing design is the processing power and the development time necessary to complete design customisation.

Hardware platforms exist now that combine powerful processors with integrated modelling, development and coding tools. The table below summarises the potential improvements faster hardware will enable.

Feature	***Existing Design***	***Next generation***
Controller strategy	*Fixed P or PI control (I has limited adjustment)*	*Fully adjustable P, PI or PID*
Controller term adjustment	*P fixed by component values. Optional I action rate and limit adjustable. Manual optimisation/selection.*	*Fully Tuneable PID with in built settings optimiser.*
Non linear amplifier	*Gain reducer effectively working in powers of two. Adjustable X breakpoint.*	*Fully configurable two dimensional look-up table.*
Servo Compensator	*Fixed general purpose RC network. Some applications require additional/alternate external to the controller.*	*Accomplished by appropriate choice of PID setting.*
Constant OLVT closure time	*Rate auto ranges in a known time band. (30-60 secs say)*	*Rate dependent on initial valve opening*
On line tuning	*Settings held in non volatile memory (E2) therefore adjustments made only when the controller is off line.*	*On line optimisation.*
Lift-flow linearisation	*Currently in the TMR channels*	*Implementation of this function in the valve controller.*

6.7 Existing fault detector extent

A steam valve scheme employing dual feedback loop with a high pressure servo would detect:-

- Any single demand fault (loss of signal, median error or signal corruption)
- Two out of three demand faults (as above)
- Servo-valve fault (wiring, contamination or coil fault)
- Positioning Error fault (valve not matching demand)
- Pilot ram feedback signal fault (faulty LVDT, driver, wiring, mechanical damage)
- Main ram feedback signal fault (faulty Resolver, driver, wiring, mechanical damage)

6.8 Improvements to the fault detector

The fault detector is significantly more complex than the controller as it includes the MMI and a large proportion of the module IO signals. The system has the opposing requirements of minimising nuisance valve trips to increase unit availability while protecting the unit by automatically closing the steam valve should a fault or combination of faults develop that may lead to prejudiced valve control.

Feature	*Existing Design*	*Next generation*
Local Maintenance status indicator	*Fault Code (requires interpretation)*	*Dedicated maintenance PC or LCD (alpha numeric messages)*
Documentation/User understanding	*Refer to the O&M manuals*	*Logic diagrams available on the maintenance port*
User Configuration	*Minimal - Program fixed in EPROM* *Few On/Off flags only*	*Maintenance PC running equivalent tiered graphical model of existing arrangement/settings.* *On line configuration.*
I/O configuration	*Fixed in application code (EPROM)*	*As above*

6.9 A Typical Valve Conversion Example

Rolls Royce were asked to propose an upgrade to a near twenty year old TG set originally supplied with flyball governing and a common low pressure pilot oil pressure control line to all steam valves. The problems the replacement system was to address were:-

- Stiction in the pilot pistons.
- Poor speed control accuracy, repeatability and stability.
- Interface upgrade to a new station DCS.
- Additional control modes.
- Unreliable unloading schemes.

The arrangement proposed was a full EHG conversion employing individual valve control. The diagrams below show, as an example, the 'before and after' changes proposed to the HP throttle valves. The remainder of the conversion to EHG involved:-

- Removal of the Main/Auxiliary governors, governor cross shaft, worm and wheel, complete pilot oil system, boiler pressure/vacuum unloaders and the pilot oil trip valve.
- Removal of the Governor/Intercept Valve relay pilot pistons, feedback cams and associated feedback linkages.
- Addition of a speed measurement toothed wheel and probe system.
- Addition of a servo actuator/position feedback monitoring set on each control valve.

6.9.1 Operation and characteristics of the original design

Operation: The unit was run to speed under control of the HP ESV's. The HP throttle valves are wide open initially, following the common pilot oil pressure signal. Near rated speed the mechanical flyball comes into range and restricts the pilot oil flow to lower the pressure in the common control oil line so closing the steam control valves. The speed control changeover point occurs when a balance of steam flow and friction results. The motor driven setpoint sets the quiescent control speed. An auxiliary governor assists the unit performance on load rejection.

Characteristics: Run to speed control is open loop (ESV's) making automation (sequenced) of the associated plant significantly more involved. The steam valve operating times were relatively slow and the valve crack points varied with temperature, due to thermal affects on the lubricating oil and linkage differential expansions. Speed control response was sluggish caused by the low inherent gain (droop) necessary for stable speed control around rated speed. Once synchronised, the load rejection performance required back-up. (A second 'Auxiliary' governor and oil dump solenoids). Valve crack points and sequencing are set mechanically as pilot oil pressures. A mechanical cam is employed to linearise the flow-lift relationship. Pilot rams are known to suffer from stiction. On Load Valve Test facilities minimal.

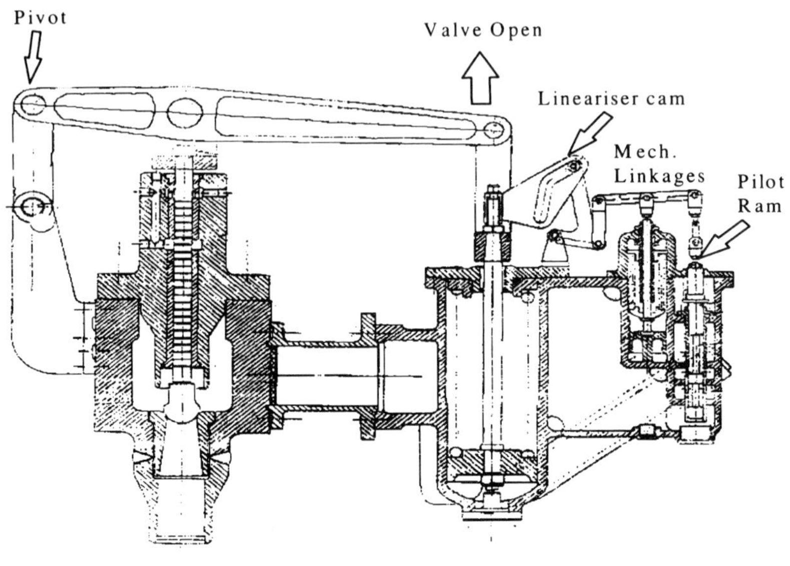

SECTIONAL ARRANGEMENT OF GOVERNOR VALVE AND OIL RELAY

**** Before Conversion**

6.9.2 Operation after conversion to electronic governing

Valve Control: Each steam throttle valve is controlled independently. Actuation is direct to the pilot ram and individual closed loop position feedback near the final valve spindle enables the position to be controlled accurately. Valve speeds are significantly increased, dead-band and thermal drifts are minimised. Each valve can be exercised on-line, electronically.

EHG Features: Speed sensing and all processing is based on a 10mS program cycle. Speed control is much tighter due to better valve control.

Run to speed can be under control of either the HP ESV's or Governing valves. Full range of EHG control modes enabling EHG-DCS systems to be seamlessly integrated. Modes are fully selectable. 'Linearisation' of the flow-lift (therefore Operator Setpoint to actual MW's) is in the controller, adjustable on load.

6.9.3 Other factors to consider…

The condition of the remaining valve actuation system components should be well maintained to make best use of the new EHG equipment.

Frequency regulation with the new system will be faster and more accurate, inevitably making more demands on the steam supply system, however the EHG can be tuned to accommodate both boiler and grid requirements.

On line changeover between full and partial arc control is feasible, after this conversion.

On any large set with multiple steam inlet paths consideration should be placed on the redundancy of the processing system adopted, the aim should be on-line fault identification and rapid repair.

The HP steam chests will need some method of 'steam warming' to remove trapped condensation prior to run-up. External mounting of the eddy current speed probes would enable on line probe replacement as well as control down to barring speed. A valve condition monitoring system (OLVT) is easily added to the basic EHG enabling valve footprints to be compared, over time. Maintenance costs can be reduced as a result.

HP Governor Valve and oil relays

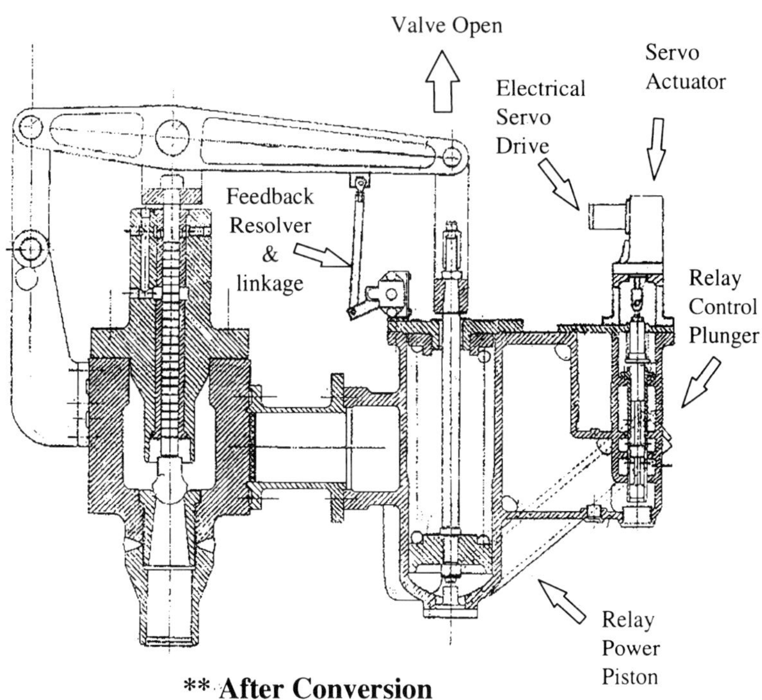

7. THE FUTURE

7.1 Software

Developing control systems using modelling techniques which incorporate models of the controlled plant is proving to be extremely successful.
When this approach is used in conjunction with 'auto code generation' the time to generate fault free software is drastically reduced.

Engineers can concentrate on the application without being burdened with the intricacies of generating and maintaining quality software.

The functionality can be proven quickly and easily using the modelling tools. This, in conjunction with fault free code generation, greatly simplifies and speeds up the equipment testing phases.

7.2 Hardware & System Architecture

In recent times there has been a trend towards integrating the EHG with the station Distributed Control System (DCS), although the extent of the *integration* is often questionable.

In the very latest designs traditional DCS systems are being displaced by systems based on Programmable Logic Controllers (PLCs).

Triplex Modular Redundant PLC architectures are available now, but the software execution times are questionable for EHG and AVR applications, and custom hardware will always be required for the more demanding input/output requirements

As PLC execution times improve and architectures become more mature it will be possible to construct acceptable EHG and AVR controllers using off the shelf PLCs with the addition of only a small amount of custom electronics.

EHGs and AVRs have a lot of common requirements with regard to measurements and interfaces to higher level control systems. In order to reduce costs and achieve the best possible availability, EHG and AVR functionality should be combined within one TMR system. It is acknowledged that customers are not requesting this level of integration at the moment but this is the logical way to progress in the future.

The writer suggests that future systems will use:-
"TMR architecture using high speed PLCs which combine the functionality of both EHG and AVR, and implemented using 'auto code generation' from proven overall system models."

S548/004/98

Initial experiences with digital control for steam turbines

T NOBLE FIMechE
Peter Brotherhood Limited, Peterborough, UK

Digital controls for steam turbines were introduced commercially during the mid 1980's. Because of their versatility compared to their analogue counterparts they quickly became established as a recognised standard of control not only for speed, but for pressure, temperature, power or any parameter connected with steam turbine control systems.

The contract for a steam turbine driven alternator at a CCGT CHP plant in Sweden was awarded to Peter Brotherhood Ltd. in 1990. This machine was installed and commissioned in 1991. A schematic diagram of the plant is shown in Figure 1. Steam for the turbine is from a peat fired boiler and the gas turbine exhaust waste heat superheater. The steam turbine exhausts into a condenser, the cooling water outlet from which provides hot water. Turbine speed is 7 000 rpm and a gearbox gives a speed reduction down to the alternator speed of 1 500 rpm. The plant cannot operate in island mode as the power output from both the gas and steam turbine driven alternators is supplied only to the Swedish grid. Power for the plant is imported from the grid. To control the steam turbine, the decision was taken to use the then relatively new Woodward 505 digital controller.

The requirement for the control system was for speed regulation when running with the generator breaker open. After synchronising with the Swedish grid, bumpless transfer to steam inlet pressure (and hence boiler pressure) control was required together with an optional limitation on the temperature of the cooling water leaving the condenser. The overall arrangement of the control system is as shown in Figure 2.

DIGITAL CONTROLLER

A block diagram of the digital controller showing the configuration to carry out the control requirement for this contract is shown in Figure 3.

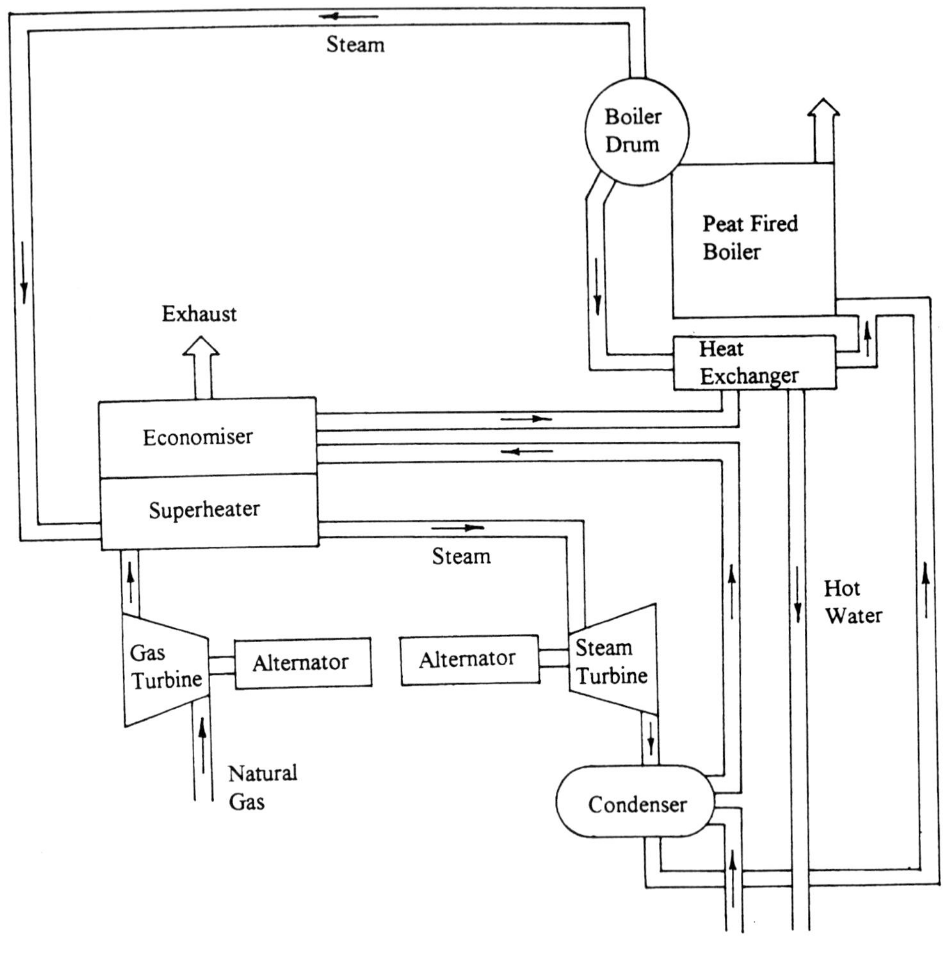

Figure 1 - District Heating Plant in Sweden

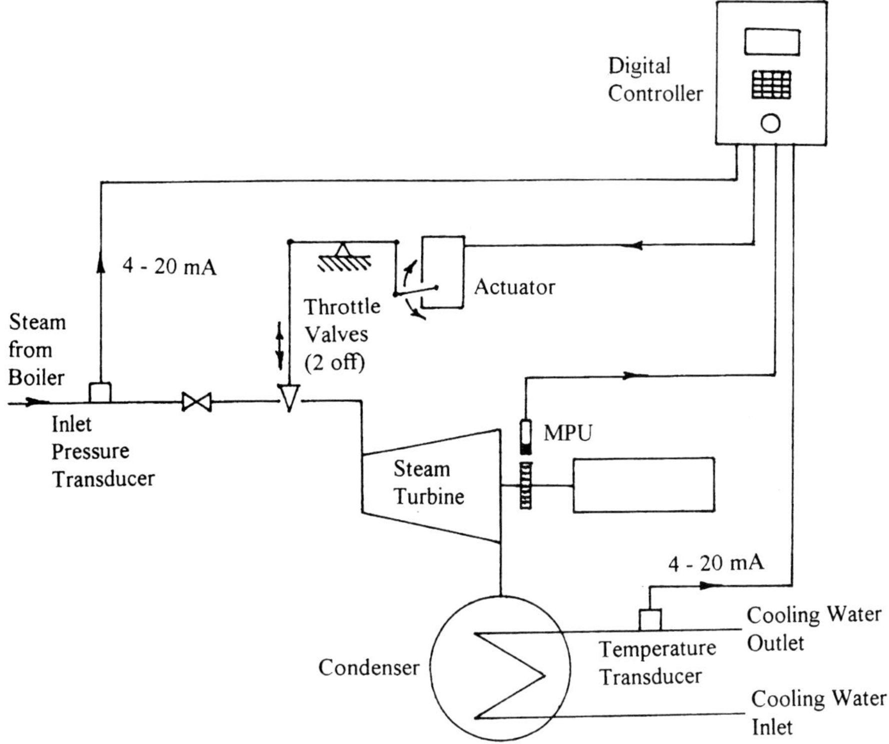

Figure 2 - Control Arrangement

It is worth considering how this particular control functions :-

On the turbine rotor is a toothed wheel. Two magnetic pick-ups (MPU's) target this gear and sense its tooth passing frequency which is directly proportional to turbine speed. This MPU frequency is then applied to the controller's speed control block where it is compared to the set point, or reference, speed and the resulting summing point PID output signal is sent to a Low Signal Select (LSS) bus. The pressure control loop is on the auxiliary control block. Turbine steam inlet pressure is sensed by a transducer and supplied to the controller as a 4–20 mA input signal. The auxiliary control loop compares the actual pressure to a set point, or reference level, and also provides a summing point PID output to the LSS. As suggested by its name, an LSS will only permit the lowest input signal to be passed to its output which, in this case, is the output to the actuator on the steam turbine. It is this actuator which controls the position of the turbine steam inlet throttle valves – see Figure 4.

For turbine start up, synchronising and initial application of load, the steam inlet pressure (boiler pressure) is arranged to always be greater than the pressure reference level. Thus for the start up and synchronising phases of operation, the output from the speed control summing point is always less positive than the output from the auxiliary (pressure control) summing point. Therefore, the LSS blocks the pressure control signal but passes the speed loop signal to

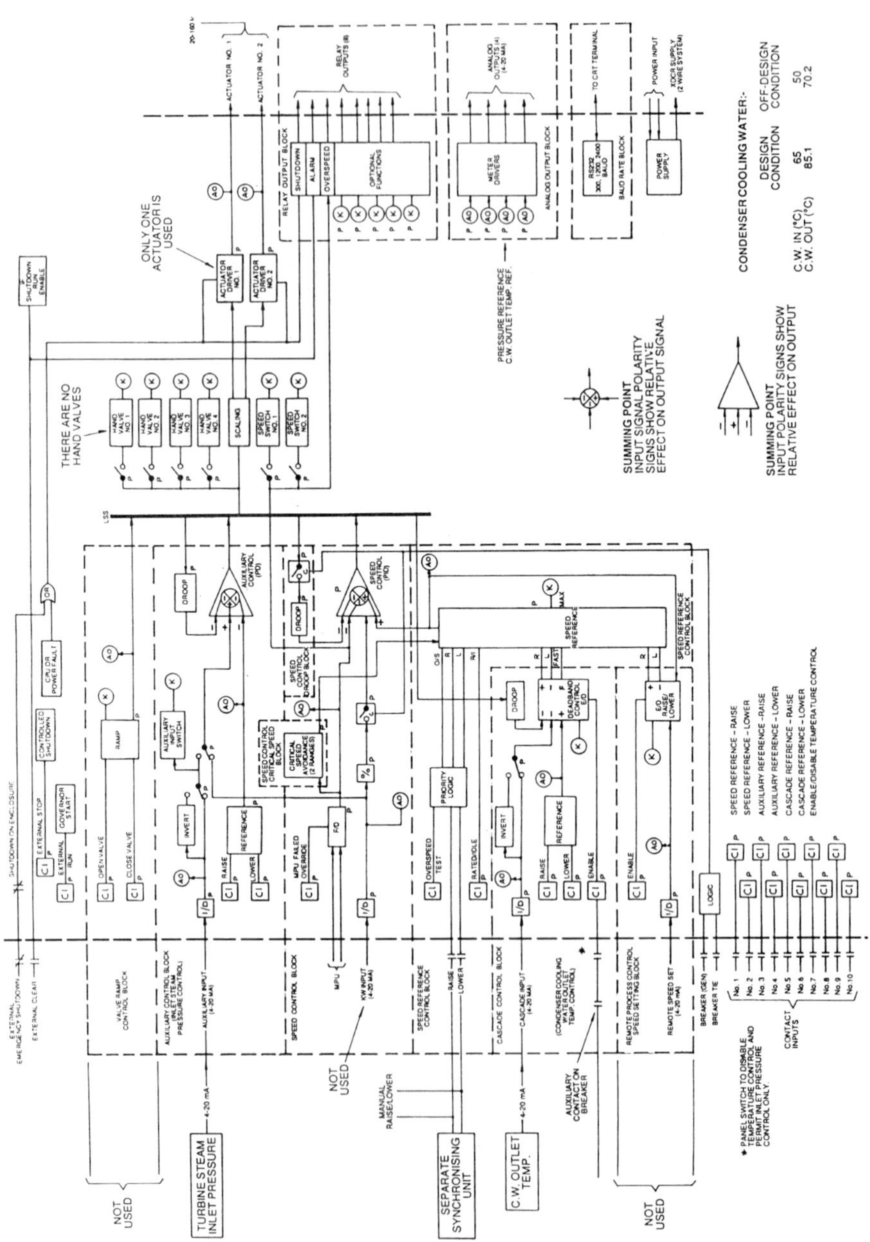

Figure 3 - Block Diagram for Digital Controller

the actuator to give speed control of the turbine. There is a small amount of droop feed-back from the LSS to the speed summing point. Once the alternator has been synchronised to the utility, the speed droop feature now permits load to be increased gradually as the speed reference is increased – load is now proportional to the speed reference set point which has effectively become a load set point. The grid now holds the machine speed constant so increasing the speed reference increases the speed control summing point output to the actuator and the alternator takes on load.

As load is increased in this manner, steam flow demand from the boiler is increased causing the steam inlet pressure to fall. This in turn causes the output from the auxiliary summing point to fall. As the falling steam inlet pressure approaches the auxiliary reference set point, the auxiliary summing point output now becomes less positive than that for the speed control summing point. The LSS now blocks the speed loop signal but passes the pressure loop signal to give bumpless transfer of control to inlet pressure. Further raising the speed reference set point now has no further influence on load and the speed set point may be raised to its maximum value to prevent interference with the pressure control loop.

If required, a 4–20 mA output signal from the condenser cooling water outlet temperature transducer can be made to influence the speed reference setting through the cascade control block. Condenser cooling water outlet temperature is limited in this way.

MACHINE FOR SWEDEN

Initial running under test bed conditions showed that when tuned, the speed control loop gave good response and was stable. These tests took the form of spin tests under no load conditions. This form of test is arguably one of the most stringent tests for steady state speed control as the turbine throttle valves are closest to their seats (see Figure 4) and hence more prone to hunting. As load could not be applied at this stage, it was not possible to test and tune the pressure (auxiliary) and temperature (cascade) loops.

Good response and stability of the speed control loop was also obtained during the initial site runs after installation but prior to closing the generator breaker.

The machine was synchronised to the Swedish grid for the first time on 19 March 1991. Immediately on breaker closure the control system hunted violently. The immediate reaction of everyone present was to press the large red emergency trip button as soon as possible!

Repeated attempts to eliminate the hunting by tuning the PID all failed: both PID's were working correctly. Before closing the breaker the control system would be stable and could not be forced into instability. However, each time the breaker was closed the control system immediately hunted violently and the symptoms were always the same:-

1) The frequency of the hunt was approximately 2 Hz.
2) The gearbox vibration levels went into the ALARM state.
3) The ammeter showed 200 amps – compared to the rated design value of 378 amps.
4) If left, the machine eventually tripped out under reverse power after approximately 5 seconds.

5) With the steam turbine hunting as described, the gas turbine would follow 180 degrees degrees out of phase with respect to the hunt cycle. This meant that the total power output from the plant to the grid was steady.

As a temporary measure, it was found possible to synchronise successfully by manually controlling the output from the valve ramp control block to the LSS (see Figure 3). Following breaker closure, load was increased by raising the valve ramp output until control passed to the pressure (auxiliary) control loop. The valve ramp output was then increased to 100 %. Once under inlet pressure control, the system controlled perfectly and was stable.

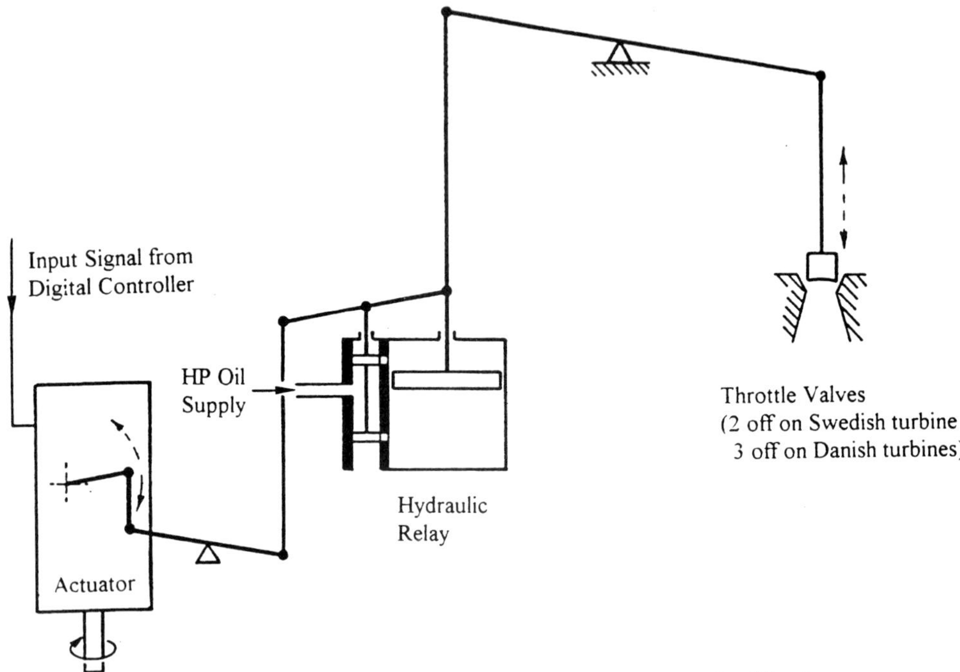

Figure 4 - Arrangement for Throttle Valve Positioning

MACHINES IN DENMARK

Shortly after the control problems occurred on the Swedish machine, two more steam turbine driven alternator sets were installed at waste disposal incineration sites in Denmark. These turbines have a rated speed of 13 500 rpm and exhaust to a condenser, the cooling water of which provides hot water for district heating – similar to the arrangement on the Swedish machine. On these machines, the control systems were similar to that on the Swedish machine except that condenser cooling water outlet temperature control was not used (just speed and pressure control only). Also, the plants in Denmark can each run in island mode. On both the Danish sets, the control systems were stable in initial running trials but when synchronised to

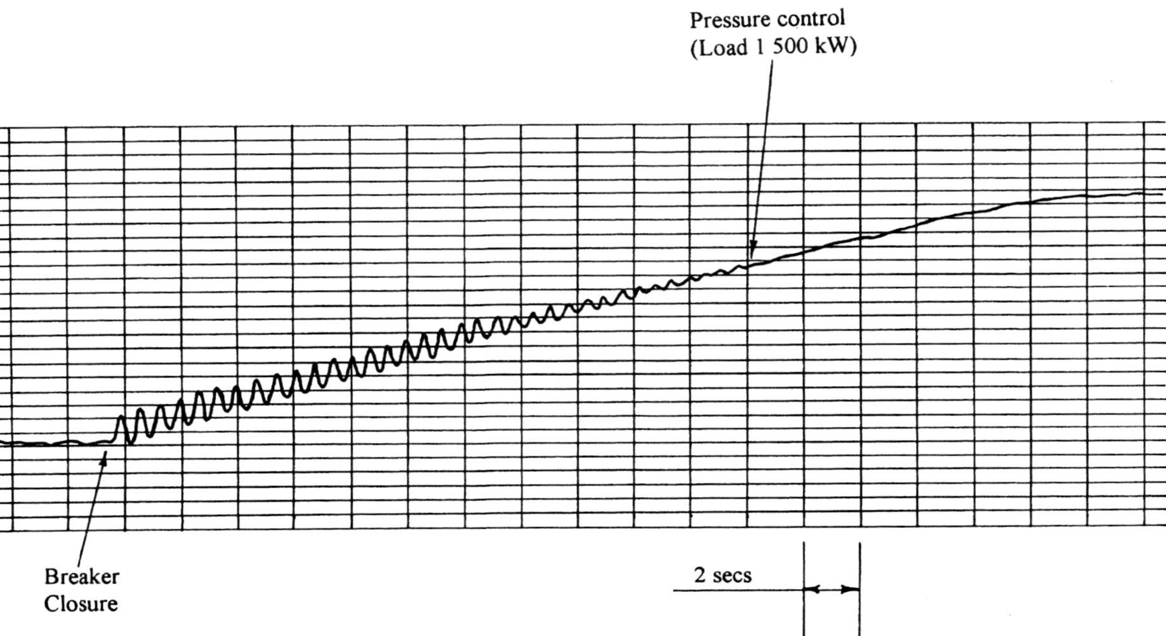

Note :- Actuator current is proportional to actuator position and to power.

Figure 5 - Actuator Current after Synchronising

the public utility immediately started hunting. However, the hunt was less severe than on the Swedish machine and allowed the alternator to be synchronised and load to be increased as originally intended and transfer bumplessly to inlet pressure control where each system then became stable – see graph in Figure 5. The hunt frequency was 1.5–2.0 Hz.

It was found possible to minimise the amplitude of the hunt by reducing the proportion gain term of the speed control loop, but this gave poor transient response – for example when each plant was tripped from export condition to island mode, the steam turbine tripped on overspeed.

INVESTIGATION OF PROBLEM

Because at the time, the use of digital controllers was relatively new to Peter Brotherhood steam turbines, the initial reaction of each of the customers was to blame the "new" control systems.

On all the machines which exhibited hunting of the speed control loop after synchronising, the PID functions were demonstrated to be performing properly. With each machine at synchronous speed but with its breaker open, the proportional gain could be increased to eventually cause instability which could then be tuned out by increasing the reset (integral gain).

The situation was initially completely baffling – why should an inherently stable system exhibit unstable symptoms when first connected to the grid? The initial thoughts were:-

a) There was an unexplained fault concealed within the control system.
b) The unexplained fault was external to the control system but was manifesting itself through the speed control loop.
c) The hunt was due to torsional vibration – each machine had a torsionally soft rubber block coupling between its speed reduction gearbox and the alternator.

In all cases the magnitude of the oscillations was found to be proportional to the proportional gain setting of the speed control loop. Also, the frequency of oscillation was approximately 2 Hz. and was found to be independent of the dynamics settings of the speed control loop. The oscillations ceased when control passed from speed to pressure control loops and on the two Danish machines, oscillations never occurred when in island mode.

It was found possible to obtain a condition of unstable equilibrium under speed loop control on any of the 3 machines under discussion while running paralleled to the grid by using the valve ramp – see Figure 3 and earlier. However, with the valve ramp influence removed, the oscillations could be induced simply by manually disturbing the throttle valve linkage as shown in Figure 4 or by changing load.

While synchronising problems were being experienced with turbo-alternator sets in Sweden and in Denmark, two other machines were being commissioned at CCGT plants in the UK These machines, like the ones in Scandinavia, both used speed control for synchronising and initial application of load and then transferred bumplessly to inlet pressure control on the controller's auxiliary block. However, on both these particular machines there were no problems with synchronising and both their control systems were perfectly stable at all stages of control.

Comparison of the UK machine with the 3 Scandinavian machines was the starting point for the investigation.

The initial phase was to identify any differences between the UK machine and the three Scandinavian machines.

It eventually transpired that the main difference between the stable and unstable machines was that the alternators on the unstable machines had fabricated pole pieces, whereas the alternators on the stable machines had cast pole pieces. According to the alternator manufacturers, cast pole pieces give higher damping values than fabricated pole pieces for torsional oscillations.

This fact gave the first important indication that the amount of torsional damping at the alternator air gap was a major influence on the oscillating condition of a turbo-alternator tied to a grid while still under the influence of a speed control loop.

ANALYSIS WORK

An alternator running in parallel with an infinite grid system has an inherent electro-magnetic synchronising torque whenever the alternator rotor departs from its normal angular position with that grid. This torque acts as a torsional coupling between the alternator and grid acting in the direction necessary to restore the original state of phase equilibrium with the grid. To draw a mechanical analogy, the restoring torque acts as a linear torsional stiffness – the torque value is directly proportional to the phase displacement of the alternator rotor from its equilibrium position with the grid.

i.e. Torsional stiffness of electro-magnetic coupling between alternator and grid, (q)

$$= \frac{\text{Restoring torque}}{\text{Phase Displacement}} = \text{a constant}$$

Taking into account the combined inertia of the turbine rotor, speed reduction gears and alternator rotor (I); plus the torsional damping coefficient at the alternator air gap (c); the mechanical analogy for a turbo-alternator in parallel with an infinite grid may be viewed as a torsional single degree of freedom system as shown in Figure 6.

From Figure 6, the general equation of motion for the single degree of freedom system may be expressed as :-

$$I\ddot{\theta} + c\dot{\theta} + q\theta = T\sin(\omega t)$$

The classical solution to the above differential equation gives the following equations :-

Undamped natural frequency, $\omega_n = \sqrt{\frac{q}{I}}$ (rads/sec.) giving frequency $f_n = \frac{1}{2\pi} \sqrt{\frac{q}{I}}$ (Hz.)

Damping factor, $\zeta = \frac{c}{2\sqrt{q\,I}}$ (dimensionless)

Damped natural frequency, $\omega_d = \omega_n \sqrt{1 - \zeta^2}$ (rads/sec.)

giving frequency $f_d = \frac{1}{2\,\pi}\,\omega_d$ (Hz.)

After obtaining the relevant values of q, c and I for the Swedish and 2 Danish turbo-alternator sets, their values for f_d were found to lie between 1.8 and 2.0 Hz. – corresponding to the oscillation frequencies observed on the machines. This fact boosted confidence in the investigation pattern. However, for all these cases, the value for the damping factor ζ was found to lie between 0.02 and 0.04. On the UK machines where the speed loop remained stable after synchronising, the corresponding values of ζ were found to be between 0.3 and 0.4 – 10 times higher than for the machines which oscillated.

Before synchronising. oscillation is not possible as there is no restoring torque, but from the instant of synchronising conditions exist for it to occur. Using these facts, it was possible to visualise what was happening on the unstable machines :-
At the instant of synchronising it is unlikely that the phase of the alternator exactly matched that of the grid. A tolerance of +/- 5 degrees (electrical) is acceptable. When the generator breaker closes, the alternator is forced to move into exact phase alignment with the grid. For each case the alternator was a 4 pole machine. Therefore, considering an acceptable phase tolerance of 5 degrees (electrical) means that the alternator physically moves by up to 2 ½ mechanical degrees.

The speed reduction gearing between the turbine and alternator acts to *increase* the phase shift of the alternator when it translates to the turbine rotor. The alternator rotor, having been pulled into phase with the grid, overshoots its equilibrium position due to insufficient damping. So now the grid exerts its restoring force in the opposite direction and the rotor overshoots equilibrium again and the cycle repeats.

Earlier it was stated that the turbine rotor has a toothed wheel acting as a target gear for the MPU's which read turbine speed. Therefore, with the increased phase oscillations at the turbine rotor, the controller's speed loop detects this as a speed change. The proportional gain term of the speed control loop (see Figure 3) serves to amplify the oscillation and reacts accordingly to vary the steam flow through the turbine by moving the throttle valve(s) – which now also oscillate. As the alternator is tied to the grid, steam flow is proportional to power and hence torque. As the oscillation is cyclic, this gives rise to the periodic excitation torque T sin (ωt) – see Figure 6. Hence, a self excited vibration is set up. The torsional motion of the shafting leads the excitation torque due to the speed loop by 90^0 (of the swing cycle) – a classical condition of resonance.

q = electro-magnetic synchronising stiffness.

c = damping coefficient at alternator air gap.

I = total inertia of alternator, gear elements and turbine rotor all referred to alternator shaft).

ω = forcing circular frequency (radians / sec).

θ = displacement from equilibrium with grid = ω t

T = magnitude of excitation torque.

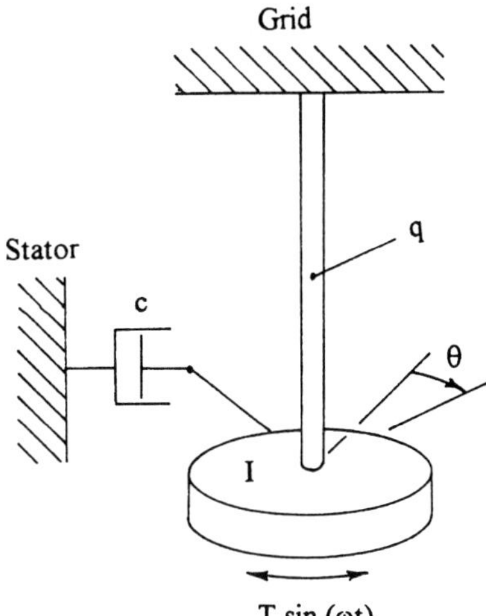

Figure 6 - Mechanical Analogy of Turbo-Alternator in parallel with Grid

Referring to the block diagram in Figure 3, when the load applied by increasing the speed reference is sufficiently high, the LSS blocks the speed loop and passes the pressure control loop instead. As the oscillations are a function of the speed loop only, the control system now stabilises. This is shown as occurring on one of the Danish machines in Figure 5. Also, when the turbo-alternator is in island mode, oscillations cannot occur as the alternator is not subjected to the electro-magnetic synchronising torque.

On systems with higher torsional damping values, the oscillatory motion decays after synchronising and the speed loop stabilises.

Returning to the turbo-alternators in Sweden and Denmark, although the oscillation mechanism was now understood, the problem now became how to introduce suitable torsional damping. To re-design and fit new pole pieces to give increased damping would have been expensive as well as time consuming. The possibility of using a hydraulic damper similar to the type used at the non-drive end of diesel engines was investigated, but this proved impractical.

FINAL SOLUTION

Having documented, analysed and eventually identified the hunting problem, it seemed that the only solution would be the costly and lengthy redesign of each alternator system. However, fortunately at the time, a new version of the 505 Control was being developed having dual dynamics on the speed control loop. Using this version with an alternator tied to the grid, changing the status of either breaker (generator or tie) changed the speed loop dynamics between pre-programmed settings. Applying this new controller to the problem machines gave the opportunity to automatically select :-

1) High proportional gain for good transient response before synchronising and for island mode operation.

or 2) Low proportional gain to minimise phase swinging against the grid after synchronising.

Arrangements were made to purchase 1 of 6 prototypes models incorporating dual dynamics which were available at the time. This new controller was fitted to one of the Danish machines and proved successful in eliminating the oscillations.

During commissioning the new controller, an unscheduled incident occurred which demonstrated its effectiveness. The turbo-alternator was at 2.8–2.9 MW which was close to its maximum rating. On leaving the machine, the author walked to the control room (a distance of 20–30 metres) where an alarm was sounding. On investigation, the alarm indicated that the tie breaker had tripped. The turbine load was now 250 kW. (the plant load) and the machine was running in island mode having successfully coped with a 90 % load rejection. Other than the alarm sounding there was no indication that the tie breaker had tripped. Had this incident occurred with the original controller, then the machine would have tripped on overspeed because the proportional gain term would have been so low in order to minimise the oscillations against the grid. Later, structured load rejection tests showed that the transient speed increase on rejecting 90 % load was approximately 7 %.

CONCLUSIONS

The hard learned lesson from the experiences with the 3 Scandinavian turbo-alternators was the fairly obvious one that the control system for a machine is affected by external influences – such as interaction with an infinite grid system. Although this conclusion was not obvious at the time.

Until using digital controls, mechanical/hydraulic type speed governors had been used for machines running in parallel with a utility. With these governor types, the ball head is usually driven through a torsionally flexible element and it is also critically damped. These features serve to smooth out torsional oscillations at the speed detecting ball head – equivalent to the speed summing point on a digital controller.

Therefore, the advent of high performance digital control systems highlighted the requirement for suitable torsional damping at the alternator air gap to provide speed loop stability when in parallel with an infinite grid.

The interaction of the proportional gain term of speed controllers with the torsional characteristics of drive trains is well known. What was not realised at the time was that the speed control loop will also interact with the torsional characteristic of the drive train with the grid.

ACKNOWLEDGEMENTS

The author wishes to thank the management of Peter Brotherhood Ltd. for encouragement in writing this paper and the staff of Woodward Governor (UK) for their assistance.

S548/005/98

A grid disconnection relay for small power producers

K F SHRIMPTON and **A G SHEARD** DPhil, BEng, CEng, FIMechE, MRAeS
Rolls-Royce plc, Bedford, UK

SYNOPSIS

A requirement of small power producers exporting to the Grid is that in the event of a Grid failure the plant should be disconnected. This is, firstly, to prevent the power generation system being damaged by re-closure of the breakers within the Grid network through the re-connection being made out of phase. Secondly, it served to prevent unearthed operation of the system.

In this paper five methods of detecting a Grid disconnection are described, and the reasons for selection of the Rate Of Change Of Frequency (ROCOF) method discussed. Development of the ROCOF relay is summarised, and the design presented in detail. The paper concludes with a consideration of the in-service experience gained to date with the ROCOF relay.

1. INTRODUCTION

There is currently a great deal of interest in the idea that minor and small scale "embedded" generation technologies could play an increasingly important role in the electricity supply industry. Although it is generally accepted that the fixed and variable cost of embedded generation plant are greater than those of larger plant, the advocates of embedded generation argue that the use of small generators close to loads can reduce transmission and distribution costs, and that this can more than offset the higher costs of generation themselves.

1.1 Embedded Generation

The advantages of embedded generation have been widely discussed, and are summarised by Driver & Ritchey (1998). The increased transmission and distribution costs associated with larger plant have been compared to the increased gas distribution costs of a number of smaller CCGT's. The smaller units were concluded to be better able to provide a lower cost of electricity to the customer, in addition to better part load performance, increased operational flexibility and reduced susceptibility to catastrophic transmission system faults.

The cost and benefits of embedded generation were studied by Marriott (1997). Novel electricity sales options were considered, aimed at maximising return on investment for plant between 10MW and 99MW. The primary conclusion of Marriott was that although installed cost per kW was relatively high, these power plants could give a good return on investment.

The practicality of small scale Combined Cycle Gas Turbine's (CCGT's) was studied by Sheard & Raine (1998). The economic viability of CCGT's with a total Net Plant Output of between 40MW and 65MW was concluded to be good when considered in terms of reduced electricity import and production of process steam.

Interest in renewable fuels continues to increase, with Mason & Sheard (1997) describing a district heating plant with a Net Plant Efficiency of 83% utilising a variety of low grade waste as fuel. That such a high Net Plant Efficiency is possible with such poor quality fuel was considered to be remarkable.

The case for embedded generation was made by Stein (1997) who concluded "cables & poles" cost over $1,000,000 per mile, with over 40% of the worlds population not connected to any utility Grid. In developing countries the block size of embedded generation plant is likely to be more appropriate to a local demand group. In developed countries, planning permission is likely to be more easily gained for an underground gas pipe line than overhead electricity transmission cables. Both these factors favour a progressive shift towards embedded generation.

The world demand for power has fluctuated over the last two decades, Figure 1. What is significant, however, is that from the mid 1980's onward the proportion of electricity produced by gas turbines, in either simple or combined cycle has increased. This constitutes a shift towards gas based power generation, which is inherently more suited to embedded generation then coal based power generation. As observed by Mason & Sheard (1997), waste to energy facilities are by nature small scale, utilising waste product from the locality. As such they are unlikely to exceed 50 MW. The progressive shift towards waste to energy facilities, therefore, also comprises a shift towards embedded generation.

1.2 Regulatory Regime for Embedded Generation

To cover the connection of minor (up to 50MW) and small (50 to 100MW) licensed generators Engineering Recommendation G.75 was prepared. The G.75 includes the procedures, information and guidance on issues such as system security, stability during network disturbances, operation without normal Grid supplies, earthing and synchronising facilities which are not normally a consideration for the smaller independent generators.

The G.75 recognises Engineering Recommendation G.59/1 and Engineering Technical Report ETR113 as sources of guidance on the technical aspects of connection. The G.59/1 states the generic requirements and ETR113 provides guide lines on how they can be met.

A key aspect of the G.59/1 is its requirement that all embedded generators should have voltage, frequency and loss of mains protection. In this paper a Rate Of Change Of Frequency (ROCOF) "Grid Disconnection" protection system is presented that meets the G.59/1 requirements for frequency protection.

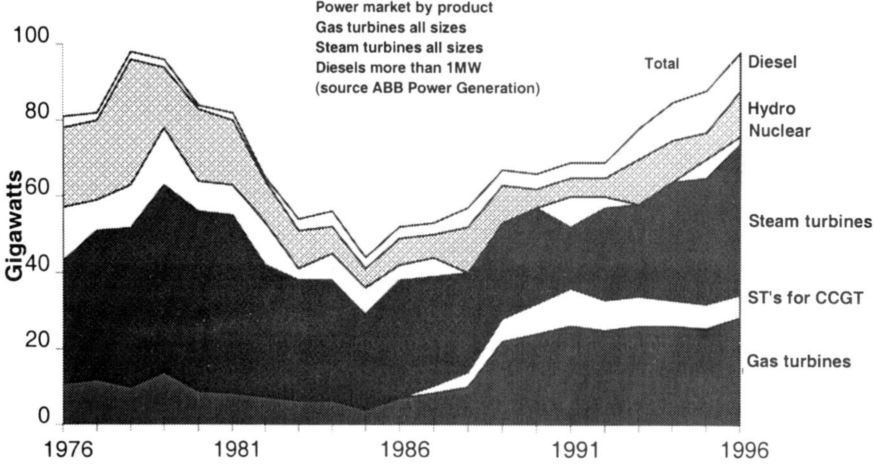

Figure 1 World electricity generation demand over the last two decades, illustrating the progressive shift towards gas based power generation.

2. GRID DISCONNECTION

The exposure of distribution networks makes them vulnerable to electrical faults. As a consequence of this, it is possible for an embedded generator to be left supplying part of a network "islanded" from normal Grid supplies, Figure 2. Islanding of a generator may leave the network without a system earth and a fault level insufficient to operate network protection devices properly. Additionally, the power quality is reduced.

The risk of islanding becomes greater when an embedded generator is capable of matching the load of the nearby network. In this instance, islanding of the generator would not in itself trip the generator, as no discontinuity of load would be detected by the generator governor. This is particularly likely to happen with embedded generators that are directly connected to a local low voltage distribution system, Figure 2. A fault that left an embedded generator servicing a local demand would result in the generator frequency drifting over time, and upon reconnection to the Grid would be out of phase.

3. SELECTION OF A GRID DISCONNECTION OPERATING PRINCIPLE

There are many methods by which a Grid disconnection can be detected. Perhaps the most fundamental of these would be to eliminate the need for disconnection by installing synchronise check relays at all points of disconnection within the Grid. This would enable islanded generators to back synchronise to the Grid, thereby eliminating the need to disconnect them. Historicaly there has never been a need for this, however, the emergence of embedded generation makes this a logical future upgrade to the Grid.

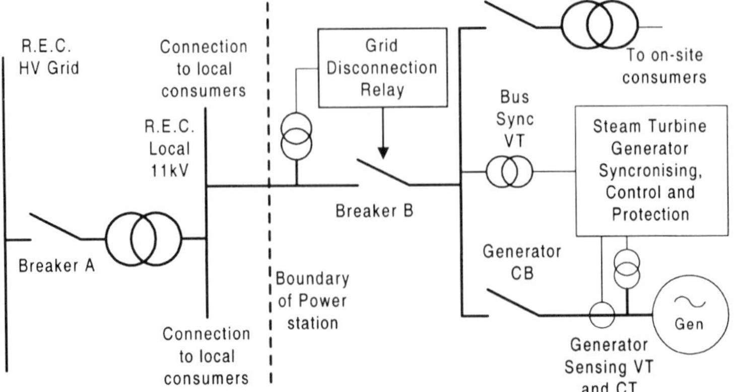

The Grid Disconnection Relay will trip Breaker B in the event of Breaker A being tripped.

Figure 2 A schematic representation of a typical embedded generator, with Grid disconnection relay fitted.

The requirement for a generator protection device was first encountered at W H Allen in 1953. At this time embedded generation was relatively uncommon, with no regulatory framework specific to it. During the 1950's the various possible Grid disconnection operating principles were studied, but never developed.

By the mid 1980's, the W H Allen factory at Bedford had a continuous diesel test program, typically providing about 1MW of electricity surplus to the site demand which was exported to the Grid. By coincidence local demand at Breaker A, Figure 2, was approximately 1MW, therefore leaving the generator exposed to islanding.

There were no regulations at the time compelling W H Allen to disconnect in the event of Grid failure. The risk of damaging the generator, however, was considered unacceptable. The decision was made to develop a Grid detection technique that would function, even with little change in load when the Grid failed.

At this time work initiated in the 1950's was revisited. Following an initial consideration of possible detection methods, five operating principles were identified. This was not an exhaustive list, as illustrated by Salman (1997) and Redfern (1997). In the following sections the five operating principles identified are presented.

3.1 Hard-Wired or Telemetry Method

In principle it is possible to identify all local public supply breakers which could disconnect a generator from the Grid. Signals from these breakers could be transmitted via a hard-wired or telemetry system and used to disconnect the generator in the event of Grid failure.

3.2 Speed Hunting Method

This method requires the governor control signal to be modulated with a relatively high frequency signal such that during normal operation the control system is stable, but following a Grid disconnection would become unstable. During normal operation the governor controls load against the Grid. Following a Grid disconnection, however, the governor reverts to speed control. In speed control the modulated governor signal would introduce a small controlled hunt into the system. The relay would detect this hunt, trip the Grid tie breaker and at the same time remove the modulating signal.

3.3 Harmonic Distortion Method

This detection method relied upon the Grid having a stable harmonic content, and the embedded generator having a significantly different harmonic content. It was theorised that it should be possible to detect a Grid disconnection by monitoring the harmonic content of the Grid, and tripping the embedded generator if it shifted suddenly from the Grid norm to the embedded generator signature.

3.4 Reactive Export Error Detection (REED) Relay

This detection method controls generator excitation via an Automatic Voltage Regulator (AVR). Reactive current in the Grid tie can, therefore, be set to a value which cannot occur naturally if the Grid becomes disconnected. The REED relay senses loss of Grid as a change in reactive current, and trips the Grid tie breaker. This operating principle was considered promising and patented, Warin (1988). In practical application it was considered that spurious trips due to natural variability of the Grid reactive current could be avoided. This would be achieved by building a time delay into the system, tripping the generator only if trip conditions persisted.

3.5 Rate Of Change Of Frequency (ROCOF) Relay

Generally, a frequency change and voltage shift will occur when a generator is islanded. The amount, or rate of change, depends on the mismatch between the generator and its trapped load. When rate of change of frequency exceeds a prescribed limit, then this can be used as a signal to trip the generator. This operating principle was considered promising, and was also the subject of a patent, Warin (1987).

Variability of the Grid load and the constant switching in and out of other generators on the Grid causes phase shifting and variations in Grid frequency. It was recognised that the associated rate of change of frequency could be greater than the limits set in the relay which would result in nuisance tripping.

3.6 Selection of the Preferred Operating Principle

The hard-wired or telemetry methods were discounted because they relied on a network of connections to public supply breakers. The cost of linking to all relevant breakers, which required the co-operation of the local distributor, was prohibitive. For these reasons the hard wired or telemetry systems were discounted.

The speed hunting method was discounted as inertia of the generator was too high to enable a suitably high frequency hunt to be introduced. This relatively high frequency was necessary to ensure that the generator did not hunt significantly when connected to the Grid.

The harmonic distortion method transpires to have been an idea before its time. Modern Digital Signal Processing (DSP) chips would have enabled this operating principle to have been implemented at moderate cost. At the time, however, the technique was discounted as the digital electronics necessary were considered to be too costly. A secondary factor in the decision was that nuisance trips due to natural variations in the Grid harmonic content were considered likely to be higher then with other possible operating principles.

The REED relay had the significant advantage over other Grid disconnection operating principles that it could detect Grid disconnection, even if there was zero change in load or significant disturbance on the Grid. Four test units were field tested. Relay interfacing with the generator AVR was found to be costly. It was also found that the on-site set up had to be carried out by skilled technicians, the cost of which was prohibitive. For this reason the REED relay was dropped.

The ROCOF operating principle was chosen for the following reasons:

- (a) It could be implemented as a stand alone unit.
- (b) Minimum interface requirements.
- (c) It could be easily checked and calibrated.
- (d) Operating parameters were easily changed.
- (e) It could be easily adapted to different power generation systems.
- (f) It could be produced at a relatively low cost.

4. ROCOF RELAY DEVELOPMENT & DESIGN

The ROCOF relay, Figure 3, was developed between 1984 and 1986. Relay technical specification is summarised in Table 1 the detail design is presented in Section 4.1.

Figure 3 The ROCOF relay, packaged as a module for rack mounting.

Table One **Rate Of Change Of Frequency Relay Specification**	
Sensitivity	Phase error selection by internal switches to ±8, ±12, ±14 and ±24 degrees. The Value of N is selected by an internal binary switch and may be set to any integer up to 255.
Standard Settings	±12 degrees and N = 6.
Sensing Supply	110Vac or 230Vac +10%, -60%.
Burden:	15VA
Trip Contact Ratings:	1250VA ac, or 1250W dc within the limits of 5A.
Reset Time After Energising	5 seconds at the standard setting.
Flag Reset	Manual.
Tripping Relay	Reyrolle Type AR.
Case	Reyrolle REYMOS R6.
Mounting	Rear flange mounted or panel flush mounted.
Terminals	External screw type at rear of case.
Maximum Ambient Temp.	50°C
Insulation	2kV 50Hz rms for 1 minute between any case terminal and earth, between case terminals of independent circuits and between contacts to earth and to the coil. 1kV 50Hz rms for one minute across normally open contacts.
EMC Immunity Tests	IEC801 Part 1; part 2; Part 3 and part 4.
EMC Emission Tests	EN55011:1991 Group 1 Class B (CISPR 11:1993).

4.1 The ROCOF Relay Design

The ROCOF relay operating principle can best be understood with the aid of a logic diagram, Figure 4. A typical application diagram, Figure 5, is given for the sake of completeness.

The mains input reference signal, Figure 4, is initially fed into a Waveform Corruption Regulator (WCR). The WCR removes transient signals induced by capacitive or inductive arcing, lightening strikes and other spurious phenomena. The result of this waveform clean up is a "valid" 50Hz mains generated reference signal. The 50Hz reference signal is divided into sampling pulses of N cycles of mains. The value of N can be any integer from 1 to 255, but a standard setting of 6 has been adopted. The result is that each consecutive sample pulse width is equivalent to 6 cycles of mains.

The counter input frequency of 18,000 Hz represents one cycle per degree of phase at 50 Hz. One cycle of mains, therefore, is represented by a digital count of 360. The sample pulse clocks an 11 store shift register which for an N value of 6 will store a digital count of 2160 (i.e. 6 x 360) in register A. On each sampling pulse the output from the free running counter is stored in register A and the number previously in A is transferred to B, the number previously in B is transferred to C and so on.

For constant mains frequency, the number in the middle register (F) will be equal to the average of the numbers in A and K. The value of 2F - (A+K), Figure 4, represents the departure of F from the average of A and K. This departure of F is proportional to the rate of change of frequency, df/dt. If the departure is greater than 12 degrees, this equates to a df/dt of greater than 0.093 Hz/sec. This rate of change of frequency is the normal df/dt trip threshold.

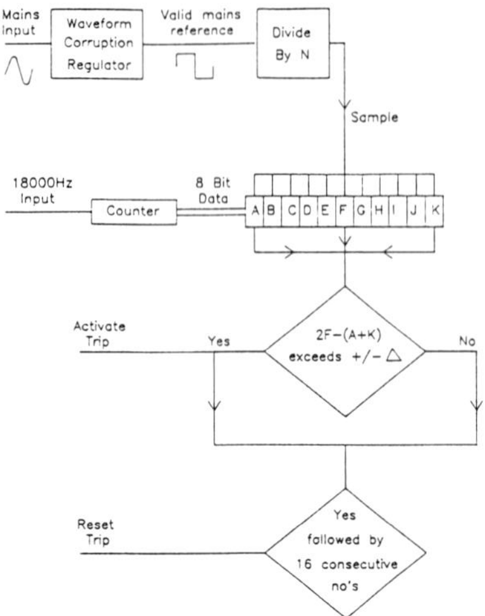

Figure 4 A logic diagram of the ROCOF relay operating principle.

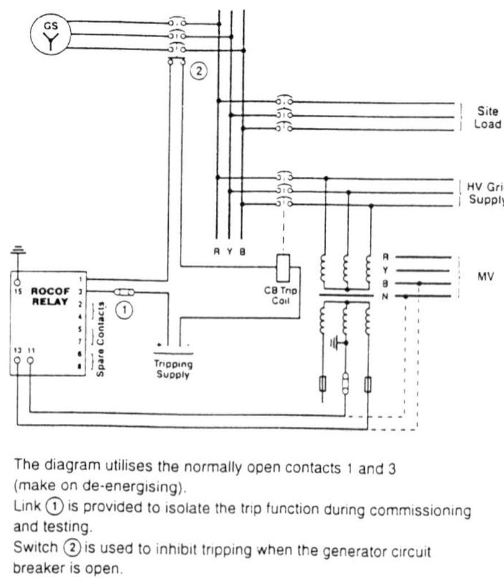

Figure 5 An application diagram of a typical ROCOF relay installation.

5. IN SERVICE EXPERIENCE

The standard df/dt setting has been found, on some sites, to result in an unacceptable number of spurious trips due to disturbance on the Grid. By changing the phase angle setting to 16 degrees (equivalent to a df/dt of 0.124 Hz/sec), the incidence of nuisance trips has been reduced. This is the standard setting adopted by electricity supply companies operating with a less stable Grid. A long transmission line between the generator and Grid can also be a cause of nuisance tripping. Under this circumstance it has been found that a df/dt of 0.186 Hz/sec minimises the instance of nuisance trips.

Extreme events on the Grid will result in the relay tripping. An example of this was the occasion on which the English Channel supply link from France failed. As a result of this several U.K. generators fitted with the ROCOF relay tripped. Trips may also occur due to voltage depressions below 30%, caused by incorrectly cleared short circuit faults. Tripping under these conditions is in fact advisable. The possibility of asynchronous operation of the generator (pole slipping) introduces the danger of damage occurring when the fault is eventually cleared and the voltage recovers to normal.

At the time of writing over 550 ROCOF relays are in operation. The operating principle has established itself as an industry "norm" and is now in routine and trouble free service.

6. CONCLUSIONS

(1) The G.59/1 Engineering Recommendation makes disconnection of an embedded generator from the Grid mandatory in the event of a Grid failure.

(2) Five different operating principles have been considered, any one of which could in principle be used to disconnect a generator from the Grid in the event of Grid failure.

(3) The rationale for choosing the ROCOF operating principle has been given, and the program of work summarised which was undertaken to develop it.

(4) In service operation of the ROCOF relay has been presented, and despite the operating principle being inherently susceptible to instability of the Grid, the relay has proven reliable in practical application.

(5) To date over 550 ROCOF relays have been installed, and are in routine service.

(6) The trend towards embedded generation has been considered. The advent of commercially viable small CCGT's and waste to energy schemes is considered likely to continue the trend towards embedded generation for the foreseeable future.

ACKNOWLEDGEMENTS

The work reported in this paper was funded by Rolls-Royce plc. The authors would like to thank David Beighton, General Manager - Allen Steam Turbines, Dr. Ian Ritchey, Chief

Engineer - Rolls-Royce Advanced Engineering Centre and Dr. Peter Malkin, Engineering Director - Rolls-Royce Transmission & Distribution Ltd. for their support and encouragement during the program of work reported in this paper. The assistance of Professor Jim McDonald and Graham Ault of Strathclyde University plus Robin Comber of Allen Steam Turbines is also gratefully acknowledged.

REFERENCES

Driver R.A. & Ritchey I., Issues Arising From the developing Role of Gaseous Fuels, to be Presented at the 17th Congress of the World Energy Council, Houston, USA, 13-18th September, 1998.

Engineering Recommendation G.75, Recommendations for the connection of embedded generation plant, to the Public Suppliers' distribution system above 20kV, with outputs over 5 MW, 1996.

Engineering Recommendation G.59/1, Recommendations for the connection of embedded generation plant to Public Electricity Suppliers' distribution system, 1995.

Engineering Technical Report No.113, Notes of guidance for the protection of private generating sets up to 5MW for operation in parallel with Public Electricity Suppliers' distribution systems, 1995.

Marriott J., The Cost and Benefits of Embedded Generation, Proc. of the I.Mech.E. Engineering For Profit From Waste - V, pp 19-27, London, UK, Nov. 11-12, 1997.

Mason R.N. & Sheard A.G., The Application of Steam Turbine Plant in Modern Waste to Energy Facilities, Proc. of the I.Mech.E. Engineering For Profit From Waste - V, pp 39-53, London, UK, Nov. 11-12, 1997.

Redfern M.A., A New Loss of Grid Protection Based on Power Measurements, University of Bath DPSP, 1997.

Salman S.K., Detection of Embedded Generator Slanting Condition Using Elliptical Trajectory Techniques, Robert Gordon University DPSP, 1997.

Sheard A.G. & Raine M.J., The Combined Cycle Application of Aeroderivative Gas Turbines, to be presented at the International Joint Power Generation Conference, Baltimore, USA, 24-26th August, 1998.

Stein H., Distributed Generation's Time May Be Here, Turbomachinery International, Vol.38 No.6, pp 21-22, 1997.

Warin J.W., A Rate Of Change Of Frequency Detection in Electricity Supplies, U.K. Patent No. 2,159,963B, 1987.

Warin J.W., An Alternator Control, U.K. Patent No. 2,159,673B, 1988.

S548/006/98

The optimization of turbine overspeed protection system test regimes

C J WATERMAN BSc, CEng, MIMechE, MIQA
Nuclear Electric Limited, Gloucester, UK
M PRINCE BEng, CEng, MIMechE
Nuclear Electric Limited, Bridgwater, UK
S R JENKINS BSc, CEng, MIMechE
ASTEC Services Limited, Congleton, UK

1. INTRODUCTION

The uncontrolled acceleration of a steam turbine to speeds significantly in excess of its design capability will lead to damage and possible disintegration. For this reason, protective systems are fitted which operate when a pre-determined excess speed level is reached and trip the turbine to terminate the overspeed event. In view of the significant safety and commercial consequences of turbine damage and disintegration, the reliability of such protective systems is of vital importance. Accordingly, it is important for the engineers responsible for maintenance and operation of the plant to understand the issues which affect protection systems' availability and performance, to be able to optimise for commercial benefit without compromising safety.

This paper describes the development and preliminary use of a reliability model representing a modern overspeed protection system for large turbine-generators. The model (known as TOPS) provides a means for assessing how the reliability of an overspeed protection system would be changed by modifications to the system itself or the way it is tested. In particular, the model could be used to support proposals to extend the period between physical overspeed tests with potential for increasing the operating availability of turbines and reducing incidence of undesirable stress.

TOPS can be used to improve the objectivity with which test periods may be optimised by enabling alternatives to be compared. It provides a valuable aid to engineering judgement.

2. OVERSPEED PROTECTION SYSTEMS DESCRIPTION

To understand how the reliability of the turbine overspeed protection system can be assessed, it is first necessary to consider how it works.

The attached Figure 1 shows the inlet valve arrangements for a modern steam turbine, and Figure 2 shows the two systems fitted to protect against turbine overspeed: the governor system and the emergency governor system. Both systems operate by detecting the increased speed of the turbine and act to close the steam admission valves.

The key components of both protection systems are listed in sequence of operation in Table 1. The horizontal divisions of the table reflect the three functional categories of each system's components, as follows:

a) Overspeed sensing
b) Trip signal generation and transmission
c) Steam valve actuation

Thus the systems provide two independent lines of protection up to the steam valve actuation system after which the hydraulic dump valves, the main hydraulic rams, linkages and steam valve spindle assemblies are common to both systems.

2.1 Operation of a Modern Governor Overspeed Trip System (Figure 2 refers)

In the event of a turbine overspeed, the rotational speed of the main turbine shaft increases. This is registered by a toothed wheel and sensor which provides a signal to the electronic governor. Excess speed is detected by the governor which then de-energises the solenoid valve fitted on the hydraulic pack of each of the steam admission valves.

To arrest the overspeed event, it is necessary to close all the steam admission valves to the turbine, (HP and IP governing and stop valves). On each steam inlet leg to the HP and IP turbine cylinders, a governing valve and a stop valve are arranged in series and act together to achieve isolation.

During normal operation, the dump valve and isolating valve associated with each steam valve are positioned to allow hydraulic fluid pressure to be maintained in the main ram of the steam valve.

In a turbine trip event, the solenoid valve is de-energised by the governor, isolating the main trip protection fluid supply while porting the fluid above the dump valve to drain. The dump valve and isolating valve are able to relax under spring force to a position which connects the hydraulic fluid in the main ram of the steam valve to the depressurised drain. Once pressure is lost in the main ram, the steam valve quickly slams shut under the force of its powerful closing spring.

2.2 Operation of a Modern Emergency Governor Overspeed Trip System (Figure 2 refers)

(For reasons of clarity, the following describes only a single path for the trip signal).

Overspeed is detected by the turbine overspeed head, shown pictorially in Figure 3. At about 10% overspeed, the out of balance force on the bolt (or ring) overcomes the force of its retaining spring, allowing it to trip an adjacent trigger and release the latch to open the pilot trip valve (Figure 2). The hydraulic fluid in chamber 'A' of the main trip valve is able to run to the depressurised drain, allowing the spring-loaded main trip valve to operate, thus connecting the protection fluid pipework to drain.

The loss of protection fluid pressure then allows the fluid above the dump valves to drain. As described in Section 2.1 above, this causes the steam valves to slam shut under the influence of their powerful springs.

2.3 On-Load Oil Injection Testing of the Overspeed Head

Modern turbines employ one of two types of overspeed head i.e. using either overspeed bolts or overspeed rings (Figures 3a & 3b). Both designs operate in a similar way in the event of a turbine overspeed in that they operate an adjacent trigger on the trip system.

Both designs can be tested without actually overspeeding or taking the set off load. This is done by selecting either the front or rear bolt/ring for testing, simultaneously isolating the associated half of the main trip valve:

- For the bolts, a supply of lubricating oil is injected under pressure into the rotating turbine shaft, through porting, to a chamber beneath the bolt on test. The force on the bolt from the pressurised oil causes the bolt to fly out and trip the isolated half of the main trip valve via the trigger and pilot trip valve.

- For the overspeed rings, the injection oil fills cavities on one side of the ring on test, displacing its centre of gravity. As a result the ring runs eccentrically and strikes the adjacent trigger. The significance of this design difference is that the trip ring's sliding surfaces are not directly lubricated during on-load oil injection, as is the case for overspeed bolts.

3. ACHIEVEMENT OF RELIABILITY

A practicable target for protection system reliability would be to allow no more than one turbine failure resulting from runaway overspeed every 100,000 years and this is typical of the risk industry will tolerate.

Having set a target, it is necessary to assess whether the overspeed protection systems are able to meet it. The chosen modelling method needs to reflect that:

- The failure event is known but the possible causes need to be identified. This implies a top-down modelling approach.

- A quantitative assessment is needed for comparison with the quantitative target.

Both these factors determined the selection of Fault Tree Analysis (FTA) as being an appropriate modelling tool (Ref.1).

Fault trees offer a simple and consistent approach to exposing the various ways in which the event of concern can arise. In this case, a runaway turbine overspeed will occur when an event which causes the turbine to overspeed occurs coincidentally with the overspeed protection system being in a failed state.

Expressed another way:

Frequency of Runaway Turbine Overspeed per Year	=	*Probability of Overspeed Initiating Event per Year*	×	*Fractional Unavailability of Protection System*	*...........................(A)*

In the above relationship, two out of three elements are already known: the acceptance limit for runaway turbine overspeed frequency which can be set in accordance with good practice; and the frequency of an overspeed event, which is derived from the following initiators:

I1: Turbine trips coincident with failure to delay tripping the generator
I2: Sudden rejections from Grid
I3: Turbine control system failures
I4: Turbine overspeed tests

Data for all of the above are available from the wider population of turbine plant.

The final element of relationship (A) is the unavailability of the turbine protection system. This parameter can be influenced by testing practice as the following relationship shows (Ref.2):

$$Unavailability = 1 - e^{-\lambda t} \quad \text{...}(B)$$

where: λ = protection system component failure rate
t = protection system component test interval

By combining relationships (A) and (B), the effect of the test regime on component unavailability can be determined and, from that, the frequency of a runaway overspeed event. In practice, the modelling is made complex by the numerous protection system components and initiating events. Note that in fault tree terminology, any one combination of events which can lead to the outcome of concern (or 'top event') is called a 'cut-set'.

4. DESCRIPTION OF TOPS MODEL

An existing fault-tree model for turbine runaway overspeed frequency, which had been constructed to support assumptions made for the safety case of a modern nuclear power station, was used as the starting point for the development of TOPS. Once the decision had been taken that FTA would be the basis for the model, it was recognised that installation on a computer spread-sheet would facilitate sensitivity studies. However, in order to restrict the model to a manageable size, it was limited to only the 130 cut-sets which were most significant (in terms of contribution to overall top event frequency). This was justified on the grounds that the contribution from the excluded cut-sets amounted to less than 0.01% of the total.

It must be emphasised that the objective of producing the TOPS model was to develop a practical engineering tool which could be used to provide an objective basis for judgement, rather than to accurately predict event frequencies in absolute terms. For this reason it was appropriate to maintain simplicity by incorporating the following assumptions:

- all the previous assumptions for the original FTA were still valid
- common failure rate data could be applied to different manufacturer's plant and equipment
- the reliability of overspeed rings is similar to that of overspeed bolts
- only significant differences in turbine design and trip configurations necessitate amendments to produce station specific versions of the model.

The majority of failure rate data applied within the original FTA were reproduced within TOPS, after checking from specific power station data that no change was justified.

As for the original FTA, the model incorporated the possibility of common mode failures (CMF). This relates to where equipment of the same design is included a number of times in the protection system. A common mode failure would potentially affect all similar items at the same time.

Following validation of TOPS using the input data for the original FTA, the model has since been modified to suit each individual station in order to:

a) reflect the different steam inlet configurations to each turbine cylinder, together with numbers and types of overspeed trip components

b) include hot reheat steam inlet valves. (Their omission in the original FTA was permissible because it was based on the assumption that only disruptive events were of significance and the reheater at that station contained too little steam to be of concern. Fault sequences involving hot reheat inlet valves should be included if the reheaters are large or if all events which could cause speeds in excess of the works overspeed test need to be modelled)

c) include passive failure of hydraulic pipework.

These modifications were undertaken by inspection, followed by addition or deletion of appropriate cut-sets in the spreadsheet. As a result, the final model comprises a database of generic data and a specific, linked, work sheet for each power station in which the total frequency of turbine disruptive overspeed failure is calculated from the sum of cut-set contributions.

The relative simplicity of the spread-sheet format facilitates further development of the model to investigate potential benefits from plant or equipment modifications. For example, the installation of a third independent speed sensing and trip system could be incorporated simply by adding appropriate rows in the spread sheet.

5. ANALYSIS PREREQUISITES

Prior to using the model, the current test activities at each station are reviewed and those components identified which are tested by each activity.

Consideration is then given to how the tests differ from actual overspeed trip duty. As the model can only accommodate one test period for each component, judgement has to be exercised to determine the most appropriate input test regime, incorporating only those tests which demonstrate, most representatively, the correct function of the key components. This is not entirely straight forward as a number of possibilities exist. For example:

- Where stations are fitted with overspeed rings, their test period is taken to be that of on-load oil injection testing since this is considered to be a representative test

- Where stations are fitted with overspeed bolts, their test period is taken to be that of the physical overspeed test. The reason for this is that the on-load oil injection tests cannot be considered representative of normal duty since displacement of the bolts is assisted by lubrication between the sliding surfaces (see Section 2.3). An exception is made for stations where the oil injection pressure is monitored and logged. Such trending highlights any deterioration in bolt freedom and therefore makes the on-load oil injection test more valuable.

The implication of the above judgements is that the physical overspeed test assumes a different significance at different stations. It is valuable at all stations to demonstrate that hydraulic pipework is not blocked. However, the benefit of physical overspeed testing beyond this function is not claimed for stations with overspeed rings.

Another area where the significance of individual tests varies is the steam valves. As the valve full-stroke test is chosen as the key function test for TOPS analysis, the resultant frequency of runaway overspeed is independent of the frequency of valve freedom of movement tests. Although these are regarded as valuable, they are not as good as full stroke tests in demonstrating the capability of steam inlet valves to close on demand, because they only check for a maximum of 10% valve movement.

From these suppositions, appropriate test intervals for the overspeed protection components can be identified and entered as TOPS input data to enable existing or modified test regimes for each station to be analysed. The calculated runaway overspeed frequency may then be assessed against the agreed acceptance limit, and alternative options explored.

6. PRELIMINARY ANALYSIS AND FUTURE WORK

The runaway overspeed frequencies of power station main steam turbines for current test regimes have been calculated using the TOPS model. This analysis has revealed a significant variation in results from station to station. This is due to a number of factors:

- Turbine plant differences, e.g numbers of inlet valves and their interconnections
- Overspeed protection system differences; these lead to the application of different assumptions when selecting TOPS input data
- Test regime differences.

The most significant finding from subsequent sensitivity studies is illustrated in Figure 4 (shown for a station with overspeed bolts), namely the considerably greater sensitivity of the protection systems at all the stations to changes in full-stroke valve test interval compared with changes to the physical overspeed test interval.

The above comparison of sensitivities is even more marked if the station has overspeed rings rather than bolts. In this case the curve for the physical overspeed test interval is even flatter while the curve for the full-stroke test interval remains the same. The lower gradient for the overspeed test curve arises directly from the selection of different representative tests for rings compared to bolts, as discussed in Section 5.

The results from analysing the existing station test regimes, from sensitivity studies such as the one described above and the process of validating the model, give confidence that it will meet its design objective of being a useful engineering tool to assist judgement when considering changes to turbine overspeed protection systems.

The next stage of work will be to develop, in consultation with each individual station, an optimised test regime for their overspeed protection systems, using the TOPS model to assess the impact and acceptability of any proposed changes. While exact changes are not known at this stage, it is anticipated that it will be possible to increase the interval between physical overspeed tests at some stations without compromising safety. The TOPS model will be useful in assessing whether changes to the interval of other component tests, e.g. valve stroking tests, can be used in a trade-off to achieve such an objective.

This approach is illustrated in Figure 5. The modified test regime ensures that the calculated runaway overspeed frequency is contained within an acceptance limit proposed by the authors in cognisance of best practice. The new regime permits a 50% increase in the period between physical overspeed tests. However, such a regime should only be approached incrementally because the existing good reliability may in part be a consequence of regular testing.

7. CONCLUSION

The TOPS model permits objective assessment of alternative test regimes for turbine overspeed protection system reliability. Use of the model should enable commercial and operational benefits to be realised. The results may be of value in supporting proposals to increase periods between physical overspeed tests or to evaluate the implications of required dispensations arising from temporary plant conditions. The model also provides a basis by which proposed hardware modifications to increase system reliability may be assessed.

REFERENCES

(1)VESELEY W.E., Reliability techniques used in Rasmussen study, reliability and fault tree analysis, theoretical and safety assessment. Siam, Philadelphia, 1975.

(2)GREEN A.E. and BOURNE A.J., Reliability technology, John Wiley & Sons Limited, 1972.

ACKNOWLEDGEMENTS

The authors appreciate the support and permission to publish this paper given by the Director of Engineering, Nuclear Electric Limited, Barnwood which is part of British Energy.

Table 1: Protection System Components

<table>
<tr><th colspan="2"></th><th colspan="2">OVERSPEED PROTECTION SYSTEM</th></tr>
<tr><th colspan="2"></th><th>Governor</th><th>Emergency Governor</th></tr>
<tr><td rowspan="8">COMPONENT ROLE</td><td>Overspeed sensing</td><td>Toothed wheel and sensor probe</td><td>Overspeed bolts (or rings)</td></tr>
<tr><td rowspan="4">Trip signal generation and transmission</td><td rowspan="4">Electronic governor</td><td>Triggers and levers</td></tr>
<tr><td>Pilot trip valve</td></tr>
<tr><td>Main trip valve</td></tr>
<tr><td>Protection fluid pipework</td></tr>
<tr><td rowspan="3">Steam valve actuation</td><td>Solenoid valve</td><td></td></tr>
<tr><td colspan="2">Dump valves and isolating valves</td></tr>
<tr><td colspan="2">Main hydraulic rams, linkages and steam valve spindles</td></tr>
</table>

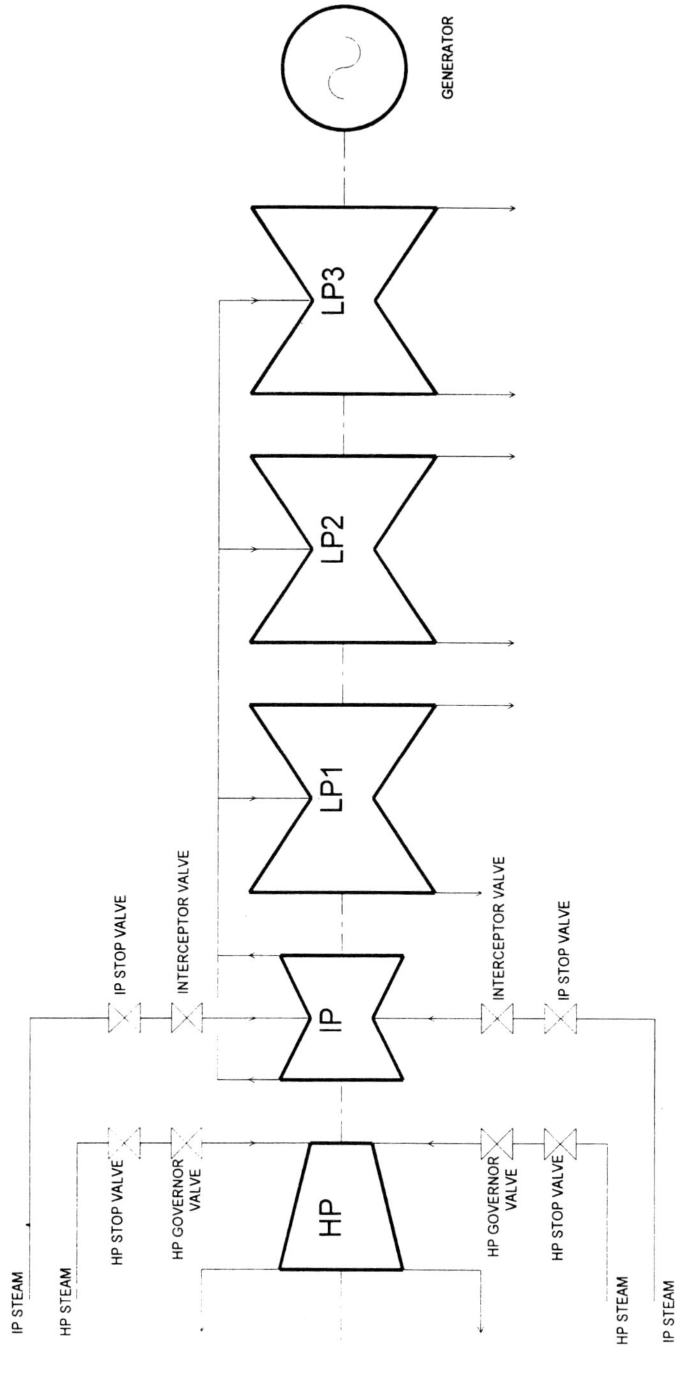

Figure 1: Simplified Arrangement of Typical Turbo-Generator Steam Supply Paths

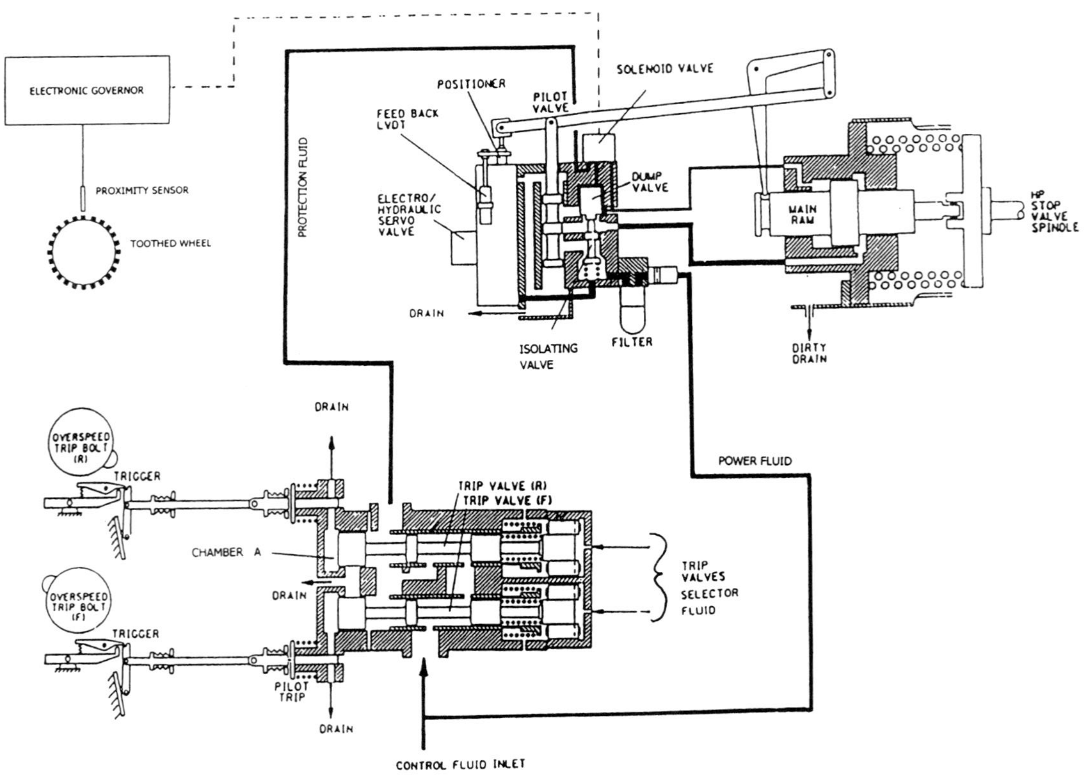

Figure 2: Simplified Arrangement of Turbine Governor and Emergency Governor Overspeed Trip Systems

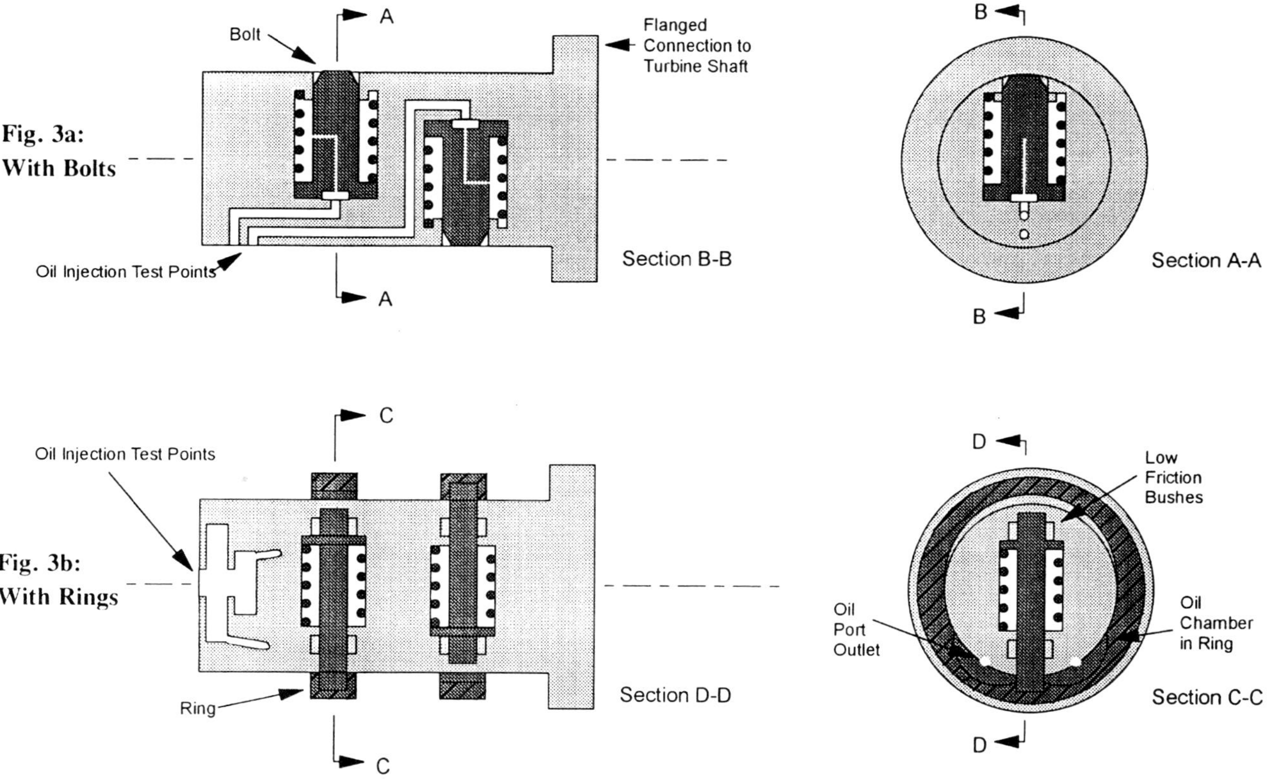

Figure 3: Simplified Sectional Views of Two Different Overspeed Head Designs

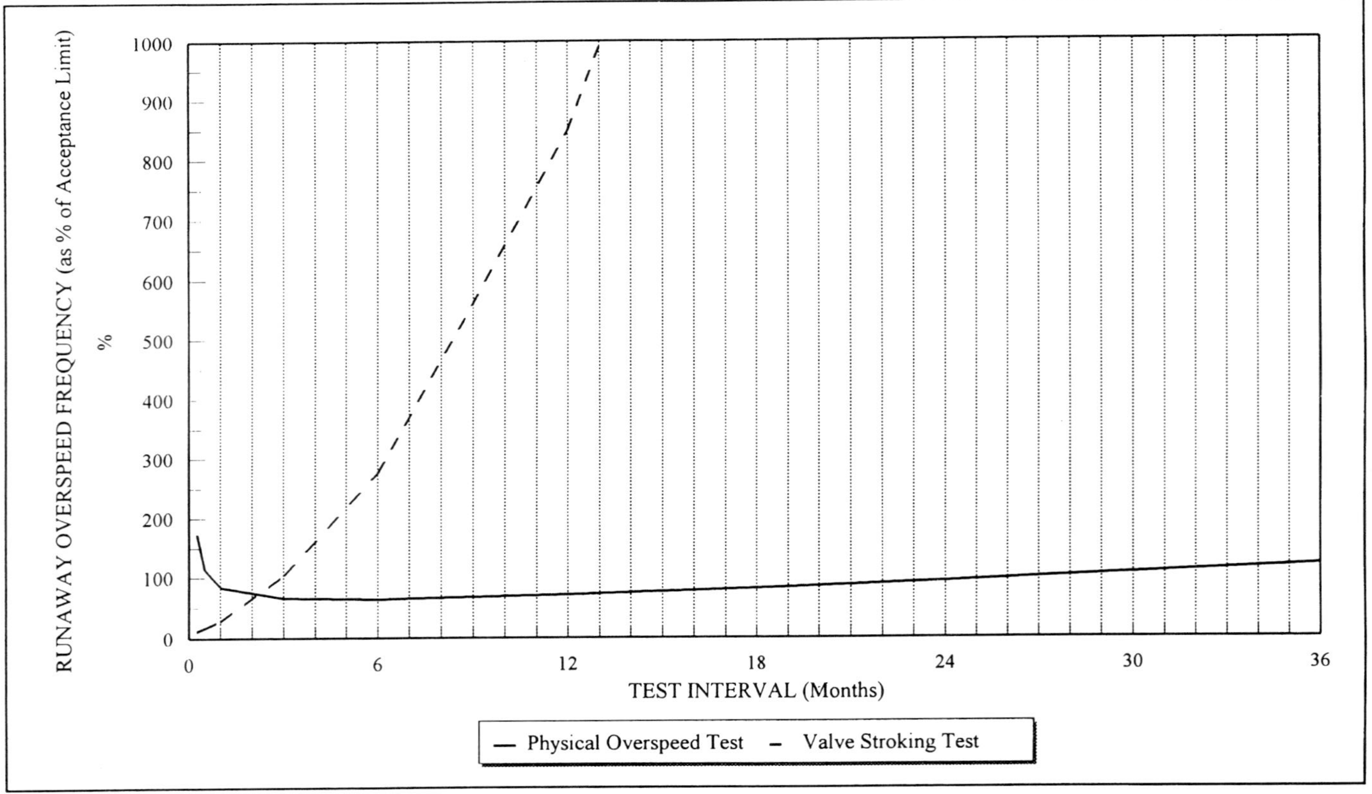

Figure 4: Comparison of the Sensitivity of Physical Overspeed Tests and Full-Stroke Valve Tests to Changes in Test Interval

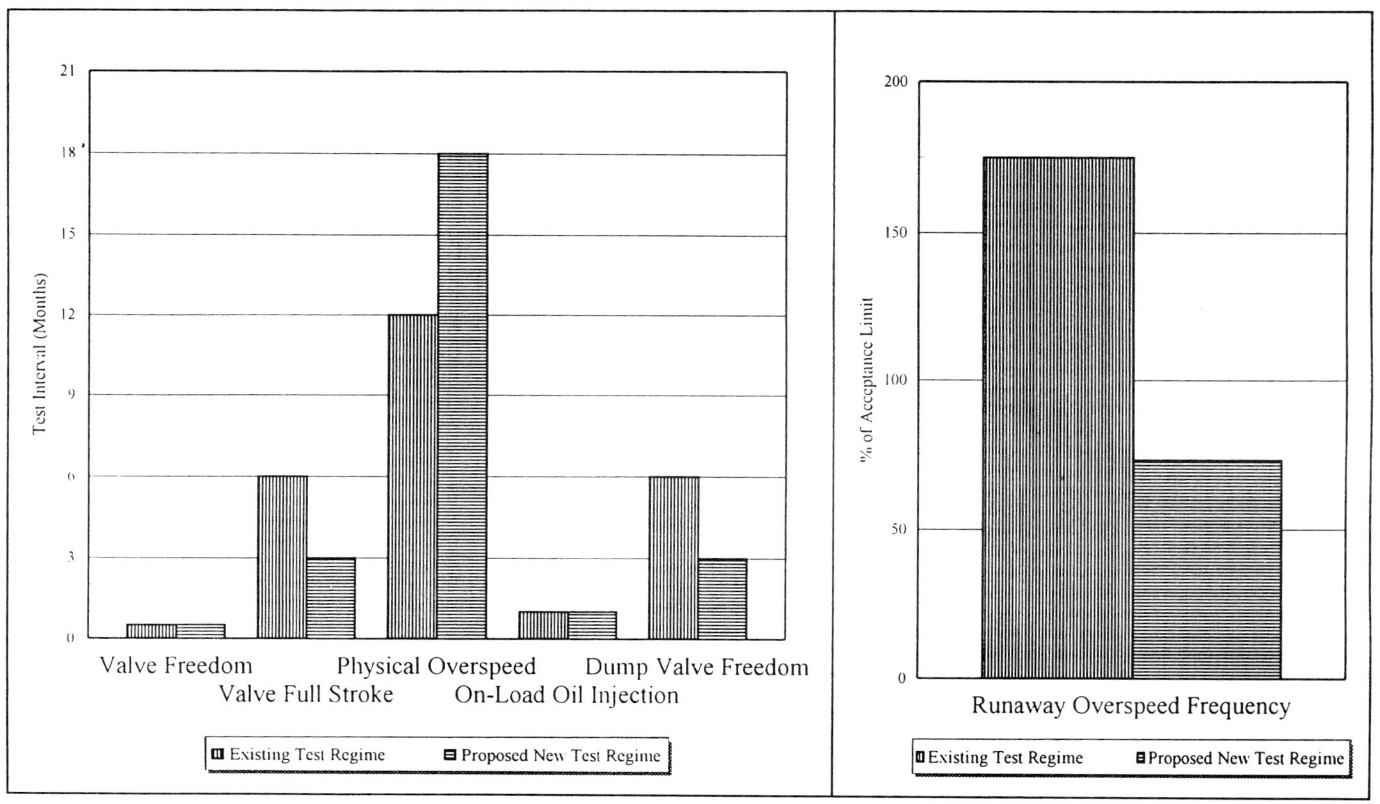

Figure 5: Possible Change in Overspeed Protection System Test Regime

S548/007/98

Overspeed protection with an insurance perspective

M ROWBOTTOM BSc, PhD, MIMechE, MIEE, FIMA
Cognant Consultants Limited, Harrogate, UK

SYNOPSIS

The runaway turbine overspeed event probably represents the 'worst case' scenario for an Engineering Insurance Company. This paper describes how an Insurer would approach the assessment of a steam turbine based electricity generation plant. It goes on to look at the role played by the testing of the overspeed protection equipment, and the relevance of these tests to the requirements in a real event. The central importance of the reliable operation of the governor and emergency stop valves is highlighted. Some estimates are placed on the required reliability of these valves. Finally some views are expressed on the role of physical overspeed testing as an indicator of system reliability, and on the requirement for an objective condition based assessment of the condition of the overspeed protection.

1. INTRODUCTION

The catastrophic steam turbine overspeed event is a classic example of a low probability, high consequence event. The event is very unlikely – the probability of such an event on a specific machine is probably of the order of 10^{-5} per year, but the associated losses on a large steam turbine, including the value of lost generation, could exceed £100M. In this type of event the expected or average loss, that is the product of the probability and the cost, is of the order of £10^3, and this number is probably meaningless as an expression of the risk.

The objective of this Paper is to provide some general information about the role of Insurance in the management of the risks of these types of events, and to consider in turn how the plant owner/operator manages this specific risk through plant testing.

2. INSURANCE COVER FOR ENGINEERING RISKS

There are many events that can threaten the financial well being of the owners/operators of power generation plant. In practice the risks of such events are managed along traditional lines. The individual areas of risk are identified, together with the associated probabilities and consequences. Risks that are judged to have adverse combinations of probability and consequence are addressed. Actions are taken to reduce the risks through combinations of operator training, improvements to the installed plant and/or its protection and the generation and implementation of control procedures for the plant operation.

At the end of this process there will be residual risks, and it will be necessary to consider carefully how these should be dealt with. One possibility is to accept the risk. That is, to accept that when generation is lost from a failure, the costs fall onto the owner/operator. This may be a sound approach for a very large and financially secure organisation. However, the majority of plant owners and operators prefer to use another approach, which is to transfer some of the risks to other people. There are a number of ways in which this is done, but a commonly adopted way is the transfer of risks to an Insurer. In exchange for an agreed cash sum, the Insurer agrees to pay for the costs and consequences of specified events. In the case of Engineering Insurance Companies the events that are typically covered are the plant damage caused by the sudden and unforeseen breakdown of the covered plant, and the business interruption losses associated with the failure.

It is standard practice to divide the losses between the insurer and the insured so that the insured remains responsible for a specified amount of plant damage and business interruption loss for each event. In this way the owner/operator retains responsibility for low consequence events together with some broadly comparable amount of the costs of the high consequence events.

Clearly the Insurers themselves will have a strong interest in the other aspects of risk management being adopted by the owner/operator. It is often specified that the Insurer will take on the risks subject to an on-going process of site inspections to identify the risk management status on the ground. These inspections typically result in Inspection Reports in which the various factors are assessed and recommendations for improvements may be identified for future discussion with the Insured. In a sense these Inspection Reports can be viewed as part of the audit process that is also an integral component of the risk management process.

In these Inspections there are various scenarios that the Insurer is interested in. These include the Normal Loss Expectancy (NLE), the Probable Maximum Loss (PML) and the Maximum Foreseeable Loss (MFL). There are various definitions of these events, but broadly speaking the NLE is a typical loss scenario giving full credit to protective systems, the PML is an estimate of the maximum loss that is likely to arise in a single event, and the MFL is the worst possible loss associated with a single event, giving no credit to protective systems. The NLE and the PML provide some guidance for the underwriter on the normal costs of breakdown and on the maximum cost that is likely to occur in a single event over the plant lifetime; the MFL is the absolute worst case that could arise. The Insurer uses the scenarios to assess the risks offered against the likely premium income and to assess the requirements for re-insurance of the risks with third parties.

One result of the plant Inspection surveys is to identify a range of scenarios of possible plant failures to determine the NLE, the PML and the MFL. For electrical power generation plant with steam turbine generation the MFL scenario is normally the overspeed failure event. It is usual in the surveys to review the overspeed protection systems and their testing regime, and to establish a review of the results obtained. This is an important specific example of the 'audit' role referred to above.

3. TESTING OF OVERSPEED PROTECTION

Overspeed protection systems are conceptually straightforward; the governor has the primary role in reducing steam flow very rapidly to limit overspeeds and it is backed up by an overspeed system that will rapidly close the steam admission valves independently of the main governor should the turbine rotational speed reach a pre-set value above the operational value. The overspeed protection is normally two parallel systems, either one of which can close the steam valves. In addition there is also an electrically operated solenoid valve that is operated by selected electrical trips. The actual equipment installed to deliver this control is complicated, with many components and subject, as the other papers in this Seminar demonstrate, to continuous changes and improvements from the application of new technology.

The basic philosophy adopted is a 'fail-safe' one; normally the steam valves are held open hydraulically against spring and, possibly, steam closing forces and the overspeed or other trips remove the hydraulic pressure, allowing the valves to close. It is important to emphasise that the overspeed protection system, while independent of the governor in its initial action, still depends on a common hydraulic system and acts to operate valves in common with the governor system. Thus, for example, if the governor action is failing because of a faulty main steam valve, this will still be a problem to the overspeed protection.

Ideally for the minimisation of risks of failure it is necessary to be assured that the governor response to rotational speed changes is correct, that the governor and the emergency stop valves are free to move over their entire range and are able to close in the specified times, that the overspeed detection system is operating correctly at its set speed, that the associated hydraulic valves are tripping correctly and promptly and that the electrical solenoid valves are functioning correctly. The soundness of the basic design is important in developing this assurance, but for continuing assurance as a machine remains in service it is usual to carry out tests of the system to demonstrate correct operation in defined circumstances.

This testing can be of the complete system or of individual components. A complete system test – a physical overspeed test – has to be completed with the machine disconnected from the grid, in the case of a synchronous machine, or from the load in the case of a steam turbine used as a mechanical drive. Such tests are not fully representative of full scale emergency operation as the steam flow used is strictly restricted, one of the two lines of protection is disarmed and the turbine rotational speed is normally run up slowly relative to a full overspeed event, to avoid unnecessary increases in machine speed, should the overspeed trip not operate. Nevertheless such tests do exercise the complete system and have, in the past, revealed changes in the performance of the system that have gone undetected by individual component tests. There is the added advantage that a measurement can be obtained – the trip speeds of the overspeed protection – and this can be used as a condition monitoring parameter to indicate changes in the condition of the overspeed equipment before they compromise the correct operation of the system.

Tests of the individual components can be carried out. It is common practice in the case of mechanical overspeed bolts to provide a means of exercising these with the machine on load, with the associated protection disarmed to avoid a machine trip. These tests provide some re-assurance, but they normally use oil as the operating medium to apply a force on the bolt, which is thereby in a situation different from that when it is required in earnest. The tests typically provide an indication of bolt freedom, but these tests do not normally make any quantitative measurement on the change, if any, in the bolt condition. It is possible for the operational speed of an overspeed bolt to change, but for this to be undetected in on-load testing.

Freedom of movement tests and valve closing time tests can also be carried out although there are sometimes restrictions on the freedom of movement tests on emergency stop valves. The valve closing times are another measurable parameter available for an assessment of machine condition. In addition to the actual closing times it can be possible to obtain data on the valve position as a function of time during the closure, a further valuable measurement.

On systems with electronic overspeed detection it is possible to use test inputs of a rising speed signal to check the response of the protection.

The frequencies with which the various tests are undertaken has traditionally been a matter of judgement, balancing the various factors. More recent approaches are seeking to make a more quantitative assessment of the reliability that is derived from the tests and to use this to establish the most relevant test schedule.

4. VALVE RELIABILITY

It is clear from the previous Sections that both the governor and overspeed system depend on the correct operation of the steam valves and their associated hydraulic components when called on to operate. While individual tests can demonstrate freedom of movement and correct closure, such tests are relatively infrequent. It is possible to regard the correct operation of a steam valve system as a random event with some probability of failing to close correctly. With this assumption it is possible to assess the reliability of the overall system given the specified reliability of the valves. It is assumed that the complete system will have 'failed' if all the valves in a steam supply fail to operate correctly. As an example, three possible configurations for valves controlling the steam admission to a turbine are shown in Figure 1. Configuration a) has two emergency stop valves, marked 'E' each protecting two governor valves, marked 'G', in a parallel arrangement. Configuration b) is a similar layout, except that there is an interconnection between the two steam supplies downstream of the emergency stop valves. Such connections may be used to minimise the load restrictions that can arise during on-load valve testing. Finally configuration c) has four parallel flows each with an emergency stop valve protecting an individual governor valve.

If p_e is the probability of failure to close of the emergency stop valve and p_g is the probability of failure to close of the governor valve then the overall probabilities of the various configurations can be calculated. The results are shown in the following table:

Configuration	**Probability of failure**
a) 2 x ESV each supplying 2 x Governor valve	$1 - (1 - 2\, p_e\, p_g + p_e\, p_g{}^2)^2$
b) as a) but with cross connection downstream of ESVs	$p_e^2 + p_e\,(1 - 2\, p_e)(1 - p_g)^4$
c) 4 x Governor valves each with its own ESV	$1 - (1 - p_e\, p_g)^4$

The resulting system reliabilities, derived on the assumption that the valve failure probabilities are equal, are shown in Figure 2. It can be seen that essentially equal reliabilities are achieved with systems a) and c), but that the interconnection between the two ESVs in configuration b) would result in severely reduced reliability, for any specific valve failure probability, if the cross connection is left open in service. It is interesting to note that the additional maintenance workload introduced from the two additional valves in configuration c) does not provide any detectable improvement in the probability of failure, for a specific valve reliability, compared with configuration a).

5. DEGRADED PROTECTION

It is interesting to speculate about the conditions that could obtain on the plant when the overspeed protection is called upon to act in anger. If the trip has arisen as a result of an electrical fault on the machine it is possible that the power supplies on the site may be subject to considerable disturbance. Alternatively, the situation may arise in which a turbine hall fire prevents access to the physical trip button on the machine and, at the same time, damages the wiring associated with the remote control of the machine. This has happened on more than one occasion.

In other words, the protection of the machine is required to operate in situations where it is reasonable to accept that the events associated with the situation may degrade the protection system itself. Given this, it is reasonable to question the relevance of the performance of the protection system under tests in a more normal situation, and also to address the derived reliability derived from the results of such tests.

6. DISCUSSION

One of the more contentious issues associated with the testing of overspeed protection relates to the frequency of physical overspeed testing. On the one hand this is a demonstration of the performance of the complete system, albeit under rather idealised circumstances. On the other hand the test requires the machine to be off load with an associated loss of revenue, although if the tests are conducted on a return to service, or prior to a planned shutdown this loss should normally be small. Objections are sometimes raised on the ground that the increased stresses associated with overspeed operation are damaging to the machine. Clearly stresses are increased on overspeed, but remain well within the design envelope and only act for a short duration. However overspeed testing on a cold machine is probably the more hazardous as it is more likely that stresses will be imposed on colder, and consequently, more brittle components. If a machine is running with a known damaged component then it is possible to make a case for temporarily suspending physical overspeeds to avoid self-inflicted further damage. For machines without known defects it could be argued that it is preferable to have a failure during a controlled overspeed test with minimal power flow through the machine rather than to have an on-load failure later, at power.

In the former CEGB it was standard practice to have physical overspeed testing not less than every six months on machines with on-load overspeed test facilities. It can be argued that this was excessive, but the policy probably derived from the extreme damage and fatalities associated with an overspeed event at Uskmouth in 1956 (Lindley and Brown, 1958).

The risk of an overspeed failure on a steam turbine leading to more or less instantaneous damage is interesting to consider. It is an extremely difficult probability to quantify. Bush (1978) estimated the value for sudden turbine failures relevant to nuclear plant to be in the range 3.1 x 10^{-4} to 3.3 x 10^{-5} per turbine year. To arrive at this value he considered recorded failures, including overspeed events back to the early 1950s. Later assessments relating to overspeed events alone have suggested values of the order of 10^{-5} per turbine year. This is clearly a figure derived from postulated reliabilities of components.

If it is assumed that overspeed testing in the UK nationalised power generation industry was conducted monthly for 45 years on a population of the order of 1000 turbines then there would have been 45,000 turbine years of experience and 540,000 physical overspeed tests. In that time there were at least two large overspeed failures (Uskmouth in 1956 and Bold in 1960). This crude calculation produces a failure estimate of 4 x 10^{-5} overspeed failures per turbine year, and this estimate it is similar to Bush's value.

More recent experience on steam turbine plant, particularly for machines in the 100+ MW(e) range suggests that overspeed failures appear to be vanishingly small. It is interesting to speculate if this is a true reflection of improved protection systems and/or improved maintenance and testing leading to a reduced failure probability, or if it is a normal consequence of a stochastic process with a low probability.

In this context it is interesting to quote from Modern Power Station Practice (National Power, 1991) : "Overspeed would have serious consequences for both plant and personnel, therefore the protective arrangements have been designed to eliminate any possibility of a dangerous overspeed". This statement is consistent with recent results from large steam turbine plant, but should it really be read as indicating that the failure probability is very low indeed. From an insurance point of view it would interesting to explore this further, as the Maximum Foreseeable Loss scenario assumes the latter rather than the former interpretation.

Given that the probabilities of failure are at the 10^{-5} per turbine year level, what is the relevance of physical overspeed testing at a frequency of about one test per year? Viewed on the numbers alone such a testing frequency is far too small to identify the failure probability. Unless it can be argued that a physical test exercises some components that are not tested individually, the data suggest that the tests are providing little assurance. However on a more pragmatic level the tests demonstrate the functioning of the complete system, and have been a feature of a successful regime in avoiding major overspeed events for over 30 years.

A similar criticism about testing frequencies can be made about the valve tests. If a probability of failure to operate of 1 in 500+ is required, how many tests are required to demonstrate the achievement this. If the tests are spread out in time then the relevance of the earlier tests becomes smaller as the valve deteriorates in service. If the tests are identifying specific problems with the valves than additional tests may be required on a specific timescale.

For both types of testing it is more important that objective measurements are obtained about machine condition and used to detect any deterioration in its early stages than to expect the tests to indicate random probabilities of failure. Perhaps it would be more profitable to seek to establish the condition of the key components on a routine basis from their performance in routine operation, and to establish more quantitative measurements of component condition from those components exercised only on test? Such an approach has the combined merit that it creates no more need for testing and generates more information about the prospective performance of this key equipment.

The questions about the performance of degraded protection arising as a reasonably foreseeable consequence of the initiation event must be addressed if quantitative estimates of system reliability are required based on expected component reliabilities. In particular the use of electrical solenoids that require power to be available for tripping to occur could be questioned.

There are some unresolved questions relating to the overspeed testing of steam turbines used as mechanical drives or which have significant power offtakes driven directly from the main shaft. To obtain an overspeed in these circumstances requires a failure of the load and a coincident failure of the overspeed system. This would remain a possible MFL event in Insurance terms, but for a plant owner/operator it might be at a level of risk that would be negligible with a lower level of testing.

7. REFERENCES

Bush, S H,	1978	Nuclear Safety, Volume 19, No.6, pp 681 to 698.
Lindley, A L G and Brown, F H S	1958	Proc I Mech E Volume 172 No. 17 pp 627 to 653.
National Power International	1991	Modern Power Station Practice, Vol C 3rd Edition, Published by Pergamon Press, Oxford, UK.

8. ACKNOWLEDGEMENT

The author would like to thank Stuart Burgess and David Phipp of the Royal Sun Alliance Engineering Insurance Company for some valuable discussions and information relating to the preparation of this paper.

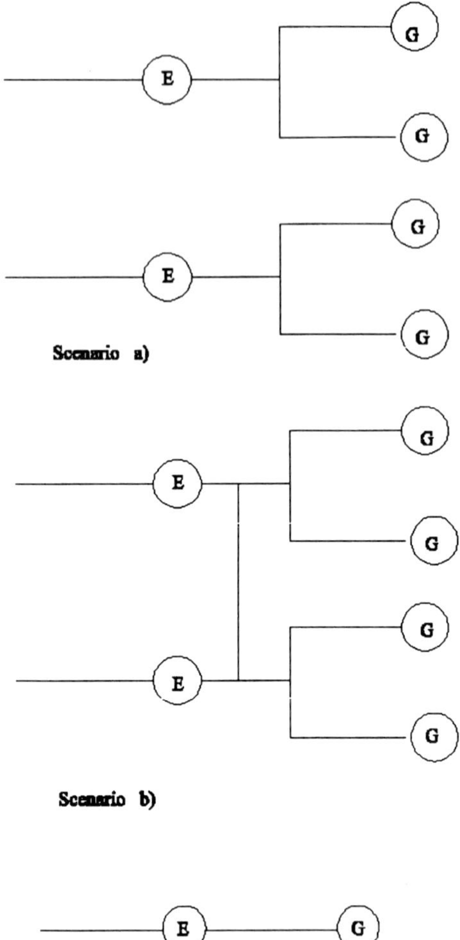

Scenario a)

Scenario b)

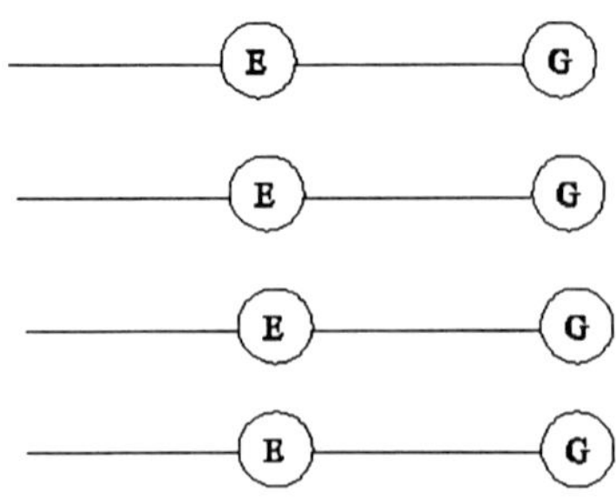

Scenario c)

Figure 1

Figure 2 System failure probabilities

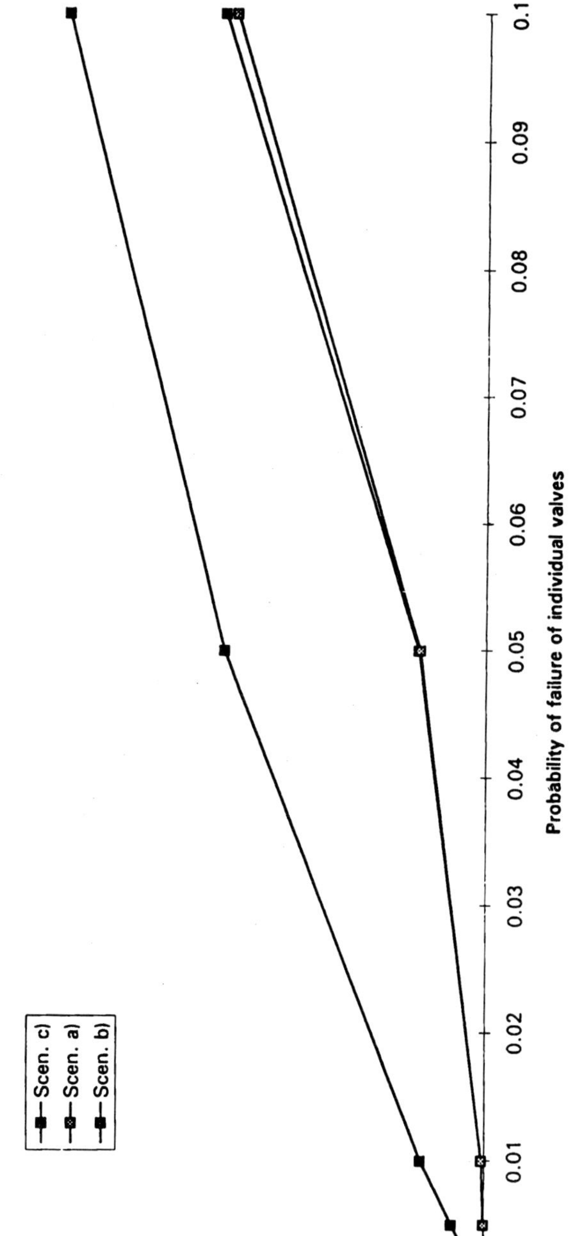

Scen. c)
Scen. a)
Scen. b)

Probability of failure of system

0.07
0.06
0.05
0.04
0.03
0.02
0.01
0

0
0.01
0.02
0.03
0.04
0.05
0.06
0.07
0.08
0.09
0.1

Probability of failure of individual valves